Nature at Our Doorstep

Matt Schuth

Nature at Our Doorstep

observing plants, birds, mammals,
and other natural phenomena
throughout the year with

Matt Schuth

University of Minnesota
Landscape Arboretum Naturalist

Nodin Press

photo: Janae Miller

Special thanks to the photographers who generously contributed images for this book, including David Brislance, Carol Knabe, John Pennoyer, Mark MacLennan, Jim Gilbert, John Toren, Sue Isaacson, D'Arcy Norman, Carol Wahl, Radim Schreiber, Normand-Watier, Joseph Yamato, Greg Thompson, Brett Laidlaw, Alex Bain, Bill Reynolds, Linnette Hulbert, Tom Sanders, Zachary Huang, Ed Book, Wayne Nicholas, Uwe Anders, Nino Barbieri, Mark Weber, Janae Miller, George Balding, and Mary Klauda.

Cover photos: Joseph Yamato (woodchuck babies), David Brislance (chickadee), Carol Knabe (skunk cabbage, fawn), Jim Gilbert (frog), John Toren (fall color).

Design: John Toren

Special thanks to the Minnesota Landscape Arboretum for allowing material that was originally published in the *Arboretum Magazine* to appear here in revised and expanded form.

ISBN: 978-1-935666-97-4

Library of Congress Cataloging-in-Publication Data
Names: Schuth, Matt.
Title: Nature at our doorstep : observing plants, birds, mammals, and other natural phenomena throughout the year with Matt Schuth, University of Minnesota Landscape Arboretum naturalist.
Description: Minneapolis, MN : Nodin Press, 2017.
Identifiers: LCCN 2017020890 | ISBN 9781935666974
Subjects: LCSH: Natural history—Outdoor books. | Nature. | Seasons.
Classification: LCC QH81 .S376 2017 | DDC 508—dc23
LC record available at https://lccn.loc.gov/2017020890

Nodin Press
5114 Cedar Lake Road
Minneapolis, MN 55416

This book is dedicated to
my brothers Robert and Mike Schuth,
and my sister, Katarina Schuth, OSF

and in memory of
my parents, Math Schuth and Marie Eversman Schuth,
and my brothers John, Ed, and Paul Schuth

Vernal witch hazel

Contents

Introduction *xi*

Spring

Winter into Spring 4
Owls 6
The White-tailed Deer 8
The Woodchuck 10
SPRING HAS ARRIVED 12
The Red-Winged Blackbird 13
Cattails 14
April Slough 16
Sandhill Cranes 18
The Forest Floor 20
Violets 21
SPRING WALK at DUSK 22
Skunk Cabbage 24
Lady Slippers 25
The Robin 26
The Minnesota Dwarf Trout Lily 28
Thawing Frogs 29
The Phoebe 30
Spring Warblers 31
Swallows 32
It's Good to Be Queen! 34

photo: Mark Weber

Aspens 35
The Chipmunk 37
Ospreys 38
Maple Syruping 40
Turkey Vulture 42
Canada Geese 44
A WALK along the ZUMBRO RIVER 46

Summer

SUMMER PRAIRIE WALK 50
Dragonflies 52
The Pocket Gopher 54
Bobolink 56
A SUMMER WALK on the NORTH STAR TRAIL 57
Cicadas 59
The Industrious Beaver 60
Cardinal Flower 62
The Goldfinch 63
Fireflies 64
Chimney Swifts 65
Duckweed 66
Touch-Me-Not 67
White Snakeroot 68
The Haunting Loon Call 69
Yarrow 70
The Ruby-throated Hummingbird 72
Snowy Tree Cricket 74

Dandelions 74
The Bald Eagle 76
WETLAND SUMMER WALK 78
Tyrant Flycatchers 79
Ed Dick's Story 82
The Mighty Oaks 84
The Fox 86
The Meadow 88
Passenger Pigeons 90
The Cowbird 92
Clouds 93
MIDSUMMER TWILIGHT WALK 94

FALL

The Wonders of Autumn 98
The Gingko Tree 100
Mushrooms 101
Wild Turkeys 103
Witch Hazel 105
Bird Migration 106
Bird Nests 108
Mother Nature in Autumn 110
Tamarack 112
Bald-faced Hornets 113
Black Walnut 114
The Red-tailed Hawk 116
The Ring-Necked Pheasant 118

Winter

A WINTER WALK 122
Cardinals 125
Winter Survival Kit 126
WINTER POND WALK 128
Watching Winter from a Window 130
Coyotes 132
The Blue Jay 133
Junipers 134
Juncos 135
The Gray Squirrel 136
Woodpeckers 138
The Horned Lark 140
Holiday Gifts 141
Snow Birds 142

Acknowledgments 145
Author Bio 147

Introduction

This book is about common things in the natural world that many people would find fascinating if they happened to notice them. Often our lives are so rushed that we can't seem to find the time for such things. And sometimes, when something interesting does catch our attention—the flutter of a bird, the sound of an insect at night—we're not quite sure what it is we've met up with. While I don't intend this book to be a field guide, I do hope that it will inspire readers to take an excursion in a park or through their own neighborhood with heightened sensitivity to the sights, sounds, and smells all around them. Several such walks are described in these pages, and there are also a number of entries, organized roughly by season, to provide a better understanding of some of the things you might see in a nearby forest, field, or marsh.

I had the good fortune to have experienced such things at an early age. I grew up on a farm near Wabasha, Minnesota, on the bluffs overlooking the Mississippi and four other rivers. Our farm was bordered on several sides by miles of forest land, so the wild things of nature were always just outside our doorstep. Birdsongs, wildflowers, and animals were our constant companions. In the spring and fall thousands of migrating waterfowl would fly over the farm, honking and calling. I remember how happy my mother was every spring when she heard the house wren singing exuberantly as it laid claim to the nest box near her vegetable garden. In the early evenings of summer, mourning doves would begin cooing in the trees behind the barn as the sun disappeared slowly over the western horizon. After the cooling showers of July, steamy fog would rise from the forests. My father said this meant the foxes were cooking in the woods again.

I suppose some of my readers have similar pleasant memories from their youth to draw upon, others not. But it's never too late to appreciate and connect to the tremendous variety of species we share this planet with. Our lives will grow richer and richer with these experiences.

Having been associated with the University of Minnesota Landscape Arboretum since 1982 has given me the opportunity to work at a variety of tasks, including managing the maple syrup production, working on wetland projects with graduate students, and helping to prepare and maintain research plots. I have set up bluebird trails, wood duck houses, and osprey nests, and recorded sightings of bird, mammal, and butterfly species for the Arboretum. I continue to write about the natural world for the *Arboretum Magazine* and do many nature walks and talks.

After one such talk in April of 2016, I was approached by Norton Stillman and John Toren, the publisher and editor of Nodin Press. They asked me if I would be interested in sharing my experiences as a naturalist by writing a book. The result of that conversation is *Nature at Our Doorstep*.

The idea of writing a book is probably the secret dream of many, but I have been motivated by the desire to share my own good fortune with others at having spent much of my life out of doors. I believe that if we fall in love with nature we will never be bored a day in our lives, and my hope is that readers of this book will be inspired themselves, or at the very least, will find it easier and more rewarding to experience the natural world in all its staggering beauty and diversity.

Nature at Our Doorstep

Spring

photo: John Toren

Winter into Spring

When the days are short, the sun hangs low, and snow covers the ground for weeks on end, even the earliest signs of spring's approach can be uplifting. Here are a few indicators that winter is relaxing its icy grip on the landscape.

- White-tailed bucks have shed their majestic racks. The chiseled gnaw marks on discarded antlers are a sign that deer mice and other determined rodents have uncovered this calcium-rich treasure trove of bone.
- The "hoo-hoo-hoo"ing of a pair of great horned owls are like love notes in the frigid arctic air. By February the owl has exerted its squatter's rights to take over an abandoned crows' nest. While a shroud of snow still covers the ground a female will already be incubating her clutch of one to three eggs.

▪ Trickling snowmelt gurgles down valleys and wooded ravines, and the strident calls of a pileated woodpecker reverberate through the trees. Its loud, persistent hammering resounds across the hills, telling us that the mating season of this crested beauty has begun.

▪ A drowsy, emaciated woodchuck has awakened from his six-month nap; amazingly, his first thoughts are of the amorous kind as he looks for a partner after his winter-long fast.

▪ As the sun's golden orb descends for the day, its final beams enflame the western horizon in colors of scarlet red and apricot orange. When the firmament darkens, the lustrous planet Venus is often the first light to appear in the west alongside the yellow sliver of a waning moon. We are one day closer to the time when Mother Nature will once again apply all of the colors in her paint box to the outdoor scene.

Hope is in the air. But nature requires patience. Spring will come.

photo: David Brislance

A half moon transforms the nocturnal landscape in early spring.

Owls

A wise old owl sat in an oak.
The more he saw the less he spoke.
The less he spoke the more he saw.
Don't you wish we could be like that wise old owl.

This bit of poetic verse presents the owl as a smart bird. Humans often picture owls with wire-rimmed glasses holding a book as if they were avian college professors. But in the real bird world, owls don't rate very high on the intelligence scale. Members of the Cordivae family—crows, ravens, blue jays, and magpies—are the real feathered geniuses.

Owls, on the other hand, are the superb terminators of their kind. These superior killing machines have eyes 100 times more sensitive to light than human eyes. Their eyes are immobile, but their vertebrate and skull are designed in such a way as to allow them to turn their head 270 degrees. The penguin is the only other bird whose eyes are set squarely in the front of its head.

In spite of these ocular advantages, owls hunt more by sound than sight. The facial disk captures sound and directs it to the ears, which are openings in the skull covered by feathers. The great horned owl's ear tufts are feathers. Some owl species can hear a mouse squeak as far as half a mile away. Research has shown that sightless owls can catch prey by sound alone. The edge of an owl's primary feathers are serrated, which breaks down turbulence and muffles the sound of air as it flows through the wings. This soundless flight makes the owl the bird version of a stealth bomber hitting unsuspecting targets silently.

Owls hunt mainly at night when their prey is most active and they have less competition from other raptors such as hawks.

Owls can kill prey much larger than themselves. Great horned owls weigh about three and a half pounds but can kill rabbits, feral cats, opossums, skunks, and raccoons. On a spring walk at the Arboretum our group saw a pair of adult turkey legs sticking out of a great horned owl nest, as if the owl family was having an early Thanksgiving dinner. Owls have been known to attack, kill, and eat wild turkeys twice their size.

photo: John Pennoyer

photo: David Brislance

The two owl species most commonly seen and heard in Minnesota are the great horned owl (left) and the barred owl (right). The barred owl has a four-note hoot that ends in a rich trill that's most easily heard at close range. It has been characterized as *who cooks for yooouu*. It's the only local owl with brown rather than yellow eyes. The great horned owl has a more complex five-note hoot.

Because of its mournful cry the owl is often associated with death. In Shakespeare's play *Macbeth*, Lady Macbeth takes the cry of an owl as a sign that her husband has killed his rival, Duncan. She states, "It was the owl that shrieked, the fatal bellman." And when the Chinese hear an owl they say "he was digging a grave."

On a lighter note, a pregnant woman in France hearing an owl screech takes it to mean she will have a baby girl.

The hushed silence of an owl belies the avian tiger within.

photo: John Pennoyer

The White-tailed Deer

Watching a regal white-tailed buck, with his massive antlers gleaming in the sunlight as he stealthily moves along a woodland hillside, sets the human heart racing. By November bucks are in full rut and searching for does ready to mate. A buck will mark his territory by thrashing trees and shrubs with his antlers and scraping the ground, all of which leave scent signs for passing does. Females that successfully mate will typically have one fawn their first year and twin fawns in the years that follow.

The picture of a fawn secretly camouflaged in a field of grass is a symbol of gentleness and peace. Bambi has become the poster child for the orphaned fawn, and *The Yearling*, Marjorie Kinnan Rawlings' heartbreaking story of a boy and his mischievous pet deer, Flag, has become a children's classic. In the real world, a fawn is nursed four or five times a day by its mother but otherwise left hidden from predators. After about a month the dappled youngsters begin to follow their mothers, and they usually are weaned by eight to ten weeks. Males stay with the does for a year and females for two years. The average life span of a deer is eight years. A buck is mature at five years and over the hill at seven.

As a deer flees from predators, its flashing white tail serves as a warning sign to others. Deer will also blow and snort to alert others and will stamp their hooves, leaving a large amount of odor from tarsal glands to alert any passing deer that danger lurks nearby.

photo: Carol Knabe

A buck's antlers are true bone and the only regenerated living tissue in the animal world. A deer can jump eight-foot fences but must be able to see what is on the other side before doing so. A six-foot wooden fence around the backyard will keep deer from invading the favorite petunia patch. The resting heartbeat of a deer is 40 to 50 beats per minute, but it can zoom to 200 in less than a second. It can go from a stationary position to 35 miles per hour in an instant. It can detect human odor a half mile downwind, and those large cottony ears can hear fingernails clicking at 75 yards. The winter coat of a deer has hollow hairs that they can puff out for additional insulation against the bitter cold weather.

The Dakota call the white-tail "tahca," which translates to "true and real meat." The largest white-tail was recorded in 1926 in Tofte, Minnesota. The deer weighed 403 pounds field-dressed and 511 pounds while living. In colonial times the exchange value of a deer hide was a dollar, hence the slang term still in use today—a "buck"—for a dollar bill.

photo: David Brislance

White-tailed deer are emblems of swiftness and grace and have adapted to our intrusion into their living space. In fact, due to changes in habitat, today there are many more deer in Minnesota than there were when settlers from Europe first arrived. Still, we should not forget who the invasive species is.

photo: Normand-Watier

The Woodchuck

Every year on February 2nd we celebrate Groundhog Day, hoping that this sleepy rodent will emerge from his hole to see a cloudy sky. Such an event means that spring will arrive early. If he sees his shadow, on the other hand, he'll head back into his hole and it will be six long weeks before spring arrives. The Pennsylvania Dutch brought this myth with them from Germany in the 18th century, eventually changing the species from the badger to the more common North American groundhog. It does offer an element of amusement, and sometimes hope, to us during a generally drab time of year.

In Minnesota we commonly refer to the groundhog as a woodchuck. It's the largest member of the squirrel family and the third-largest rodent in North America. Unlike chipmunks, skunks, and other semi-dormant animals, woodchucks are one of the few mammals that actually go into deep hibernation. During this deep sleep, which usually lasts from the end of September until early April, the woodchuck's body temperature falls as low as 38 degrees, and its breathing rate slows to once every six minutes. In preparation for this winter nap, woodchucks add half an inch of fat over their entire bodies and end up weighing up to 14 pounds. During the winter months males lose half their body weight and females more than a third.

photo: Joseph Yamato

In summer woodchucks burrow in open areas. In winter they relocate to the woods, where leaf litter can cover the openings and make it harder for predators to locate them while they sleep. A woodchuck can dig a five-foot tunnel in a single day. Its species name *monax* means the digger. Before hibernation it will line the sleeping chamber with dried grass and leaves. Females will bear four to six babies once a year. Young woodchucks are known as chucklings.

The woodchuck may not be able to predict the weather, but the Wabanaki tribes of New England consider Grandmother Woodchuck to be a wise elder with patience and understanding. We humans could take a lesson from her.

Does Groundhog Day folklore hold true in our region? In 2012, the Minnesota DNR took a look at the previous 20 years to see how well the groundhog did as a weather forecaster in the metro area. In 13 of those years, the critter would have seen his shadow. In those 13 years, the date when the snow was last one inch deep ranged from March 16 (1999) all the way to April 28th (2002). That's not much of a correlation. But speaking more generally, when the groundhog doesn't see its shadow, there must be plenty of cloud cover and moisture in the air—probably due to warm weather and melting snow. When he *does* sees his shadow, it's often due to the clear, cold air associated with an arctic high pressure zone that would be likely to support wintry conditions.

Spring Has Arrived!

Crocuses poking through the soil

photo: Mary Klauda

It's the time of year when crusty patches of snow linger in the shadows. We watch gray squirrels romp and tumble during their mating ritual carry mouthfuls of dried grass and leaves to construct their nests in anticipation of a new family. Killdeers arrive with their eponymous crescendo *killdeer, killdeer*, as if they're eager to let everyone know they've returned from their winter haunts. Chickadees raise the tenor of their songs as they sway dangling from the dried brown seed heads of coneflowers, hoping to find a few remaining nuggets within, and delicate white snowdrop flowers emerge through the leaf litter and the last crystals of snow.

Along southern exposures a crocus might poke through the soil in a race to beat the first yellow petals of a dandelion. Male red-winged blackbirds flash their brilliant epaulets as they call from wetland ponds. They're busy staking out territory in anticipation of the females, who will arrive a week or two later. A pungent smell in the air tells us a striped skunk is out and about nearby, searching for a mate. Great horned owl young peer over the edge of their nest awaiting a mouse meal from mom or dad. Little brown bats flitter out at dusk hoping to hone in on the first insects of the season.

Spring is a time for all species to inhale the freshness of the earth and enjoy the renewal to come. We humans enjoy it, too.

The Red-Winged Blackbird

Red-winged blackbirds are one of the most abundant species in North America. On their wintering grounds in the southeast there is estimated to be 150 roosts, each of which provides habitat for a million birds. This may explain why it's one of the most widely studied bird species—the ornithological "white rat."

When male red-winged blackbirds return in March from their southern homes, the wetlands resound with their *conk-la-ree* calls as they vie for territorial rights. Red-winged blackbirds have strong site fidelity; they usually mate within 30 miles of their birthplace. Their scarlet epaulets add a touch of brilliant color to the otherwise drab marshes. The bigger and gaudier the epaulets, the better chances a male will have to secure a territory. Blackbirds are polygynous, with a male having a harem of up to six females. The male that is able to defend the best nesting site will win the most female hearts.

Redwings, like other blackbirds, do a "hop, skip and jump" over each other as they feed in fields. We might also begin to skip when we hear the birds' exuberant springtime song, assuring us that warmer and greener days lie ahead.

photo: John Pennoyer

In autumn, farmers watch flocks of red-winged blackbirds descend on their corn fields and gorge themselves on the newly ripened crop. In the culture of the Plains tribes the blackbird was a symbol of corn, and the Sioux believed that if blackbirds ate their crops it was punishment for not properly honoring the corn.

photo: John Toren

Cattails

Cattails are one of the most widely distributed plants in the world. They're native to every American state except Hawaii, where they've been introduced and are now considered an invasive species. Cattails grow in shallow water. When they die they form a solid mat of compostable material in which plants that thrive in wet soil condition, such as willows and alders, find it easy to take root.

When they first appear in springtime, cattail flower heads are green. The bottom few inches of the flower hold the female blossoms, which makes it easy for the male flowers to pollinate them from above without the need for insects. As the male blossoms shed their pollen, the familiar brown hotdog-like spike appears.

Each cattail seed head contains about 125,000 seeds, which are widely dispersed by the wind. As if that weren't enough to ensure the survival of the species, cattails also spread subterraneously via rhizomes, which makes them extremely difficult to eradicate once they've become established in a marsh or lake.

Red-winged blackbirds, marsh wrens, rails, and other bird species make their nests in cattail marshes and use the dried stalks and downy seeds as nest material. Muskrats and geese feast on the underground roots, and muskrats use the stalks for constructing their houses. Mallards, hooded mergansers, wood ducks, and blue-winged teal are some of the waterfowl that use cattails for protection and cover for their young.

Minnesota's early Native American tribes used all parts of the plant. They boiled the green spikes and ate them like corn on the cob. They also collected the pollen to use as a flour supplement. They harvested the starchy rhizomes, which have as much protein as rice or maize. They wove the rushes into baskets and other containers, made baby powder and diapers from the fluff, and used the down to line their moccasins. One Native American word for the seed means "fruit for papoose's bed." Rootstocks were boiled, mashed, and then used as a poultice to treat wounds and burns.

When we're seeking a bit of solace on a wintry day, it's a pleasure to find a quiet spot where the only sound to be heard is the gentle rustling of dried brown cattail stalks on a snowy pond.

photo: Gregg Thompson

The elusive Virginia rail

April Slough

While strolling along a backwater slough I decided to rest on a white oak log and see what nature held in store. Out on the water the boisterous honking of a pair of Canada geese told me the mating season had begun, though much amorous activity still lay ahead. A single sandhill crane stood stoically among the dried brown sedges, its rust-colored feathers blending perfectly into the drab surroundings. Then I spotted its mate hunkered down on a nearby nest, incubating the family to come. Across the slough in a stand of dead American elms, a pair of northern flickers wooed each other with their *wicka, wicka* love song. The trees nearby were pockmarked with numerous holes, one of which would make an inviting home for the next brood of baby flickers.

To my right the flute-like warble of a ruby-crowned kinglet sweetened the air. How can such an exuberant, robust sound come from such a small creature? A cattail stalk, with its soft puffy seed head, held a male red-winged blackbird intent on defend-

photos: John Toren

A ruby-crowned kinglet

ing its territory. I wondered if his calls were meant to be a warning to other males or a love serenade to an interested female.

A pied-billed grebe floated by and then suddenly dove as if it had been startled by my presence. More likely it was searching for minnows and was unmindful of any human intruder. Western chorus frogs sang in rhythm, unaware that their common name has been changed to northern boreal frog. Did anyone ask them if their new name was acceptable? The low bellowing croak of the northern leopard frog rumbled across the water. Who dares to change its name! Near the shoreline a male wood duck pursued a fleeing hen. Perhaps this Adonis of the waterfowl world thinks all females should love him for his beauty.

A male wood duck in his mating plumage

photo: Carol Knabe

Sandhill Cranes

The haunting, otherworldly call of sandhill cranes flying overhead during their spring migration is a timeless avian chorus. Fossil remains of this species unearthed in Nebraska date back nine million years, making sandhill cranes the oldest living bird species on earth. Incredibly, they were almost extinct in the U.S. by 1900, due to habitat destruction and uncontrolled hunting. Fortunately, the population has rebounded, and today 500,000 sandhill cranes stop over in the Platte River country of central Nebraska every year during their flight north. It's the largest migration of any bird species in North America and includes 80 percent of the world population of cranes. The name sandhill crane derives from their migration to the sandhills of Nebraska.

In the Upper Midwest you can spot sandhill cranes in many places, though not in such large numbers, at wildlife refuges and even in fields along the interstate. Sherburne National Wildlife Refuge in Minnesota and Crex Meadows Wildlife Area near Grantsburg, Wisconsin, are two good places to start looking. If you find yourself taken by the magic of these large, graceful birds, the next logical place to visit would be the International Crane Foundation in Baraboo, Wisconsin, where all 15 of the world's surviving crane species are on display in natural settings.

Sandhill cranes have a wingspan of up to seven feet. They typically have gray plumage with fluffed tail feathers that make them appear as if they are wearing serapes. Because they forage in the mud and then preen their feathers, they develop a rusty-brown appearance. The red crown on their head is a bald spot of skin. Their exuberant, frenzied mating dance, where they jump six to eight feet in the air, throw grasses and other debris, and literally dance themselves into exhaustion, is an ancient ritual, and a must-see for all nature lovers. Sandhills nest in marshy wetland areas, where the female lays two eggs. They mate for life, with both parents protecting the young. Sandhills have a life span of between 20 and 40 years and start mating at around five years of age.

Old genealogical records were referred to as "pied de grue," French for foot of the crane, and that's where the word "pedigree" originated. The Aztecs of Mexico were from Aztlan—"land of the cranes." The Cheyenne associate them with lightning, possibly because of the red spot on their head. In Japan, red-crowned cranes are a symbol of longevity, and paper origami cranes are folded and exhibited as a protest against nuclear weapons.

photo: Carol Knabe

In flight sandhill cranes have a distinctively "hinged" silhouette, with neck and legs slightly lower than the chest and wings.

photo: Brett Laidlaw

Wild leeks, otherwise known as ramps, are worth seeking out in springtime. But harvest only a few so that the colony will continue to thrive.

The Forest Floor

The somnolence of the forest floor in spring belies the movement taking place beneath the faded bronze leaves. Patches of lily-like green leek leaves have appeared. This tangy wild onion is an herbivores delight. Because of the pungent odor of abundant leeks along the southern shore of Lake Michigan, the Miami-Illinois tribe called the place *shika'go*—skunk place. (We know it as Chicago.)

photo: Alex Bain

Scarlet cup mushrooms

The brilliant blaze color of the scarlet cup mushrooms growing from rotted twigs seems to be the work of forest gnomes with tiny paintbrushes. An empty ivory-colored snail shell is the fossil remains of its departed inhabitant. Snails never leave the shell but grow an additional ring each year. A fallen paper birch is covered with colorful bracket fungi. These fan-shaped fungi, known as polypores, come in many shades, from yellowish tan to cream, blue, ochre, and green. Other birch trees standing next to their fallen comrade have larger black and gray fungi protruding from their trunks like horses hooves.

A warmish southern wind sighs through the upper canopy as the last remaining red oak leaves tremble and hang stubbornly to the past. The seasonal shift has begun.

Violets

The naturalist John Eastman described the violet as "the low blue flame in the woods," and "the pilot light that ignites the resurrection of spring." Longfellow said about the violets, "they lurk among all of the lovely children of the shade."

photos: John Toren

Violets have five petals with the bottom one acting as an ultraviolet highway directing pollinating insects to the rich nectar in the spur of the flower. Unseen and unknown by most floral enthusiasts is the hidden flower at the base of the stem, which never opens and never blooms. It is a clone of the flowering plant and drops its seed near the parent. It is sometimes called the "summer violet" based on the time of the year when it matures. This survival mechanism ensures the violet's continued existence in case the flowering part fails to get pollinated by insects during cold spring weather.

Violets are rich in vitamin A and C and are sometimes used for making jams and jellies. The Ojibwe traditionally made a tea for bladder pain, and other tribes used it as a poultice to treat skin cancers.

Beyond their role in forest ecology, violets have also played a role in human history down through the ages. As they migrated across the steppes of Russia the Tartars survived on a soup made from the roots of violets. Monks of the Middle Ages called it the Herb of the Trinity and made a cordial from it. In exile on Elba, Napoleon promised he would return to Paris when the violets bloomed in spring. Even Nero's grave was decorated with violets each spring by unknown supporters.

Blue violets symbolize faithfulness and devotion. In Shakespeare's tragedy *Hamlet* Ophelia says, upon the death of her father Polonius, "I would give some violets, but they withered all when my father died."

Watch for these delicate blue creations of nature—they can also be purple, yellow, or white—as they play peek-a-boo through the carpet of brown leaf litter on the forest floor.

Spring Walk at Dusk

As dusk settles over a secluded pond in early spring, the sights and sounds of nature invite us in. The fading twilight is the signal for the pond's amphibian inhabitants to begin their nighttime musical.

The raspy barking call of the wood frogs joins the mellow *peep-peep* of the spring peepers. These peepers look like miniature crocodiles as their tiny heads appear and disappear on the surface of the still water. The clear, melodious tune of a song sparrow rises from the pond's edge as he perches on a spear of dried sedge. The multi-noted tune of the song sparrow resonates from a stand of scrubby junipers.

This sweet cachophony of sound might be rudely interrupted by the raucous *cao-CUCKK* call of a rooster pheasant. In the dimming light, a pair of ducks splash down into the water, visible only in silhouette. In a short while their mournful wail tells us who these new arrivals are: wood ducks.

As we move through the woods, circumnavigating the pond, a doe appears at the edge of the woods. On this windless evening, we can hear her delicate footsteps as she moves cautiously through the crisp brown prairie grasses. Her large ears are

photo: John Toren

pricked up searching for sound while her head bobs up and down seeking unfamiliar sights and smells.

A great horned owl abandons his familiar five-note *hoo-hoo* and unleashes a deranged, maniacal shrill from the depths of the forest, like the angry cry of a banshee foretelling a death. As night settles in, the American woodcock begins its aerial mating ritual. A monotonous *peent peent* call can be heard as the bird rises into the sky and descends to the ground.

On an adjoining bluff the discordant yipping of a coyote family pierces the cooling air. The outburst lasts for only a few minutes before the coyotes head off on their nocturnal hunting forays.

photo:NatureWatch

Spring peeper

Now the darkened sky opens up like an ancient black umbrella with pinholes of starlight shining through its worn surface. We head for home; the slow but unceasing movement of the seasons continues.

photo: Carol Knabe

Skunk Cabbage

The blustery gales of March can bring sleet storms that cover the trees with shimmering coats of ice. Their branches clatter together like elves tap dancing in the sky. But just as Old Man Winter begins to release his grip on the landscape, a unique plant of the marshy wetlands—the skunk cabbage—defies all logic and begins its perennial journey of life.

This solitary resident of the frozen bog (*Symplocarpus foetidus*) is a member of the Arum family and, like its cousin the jack-in-the-pulpit, has a sheath (spathe) that shrouds the flower (spadix). The spathe is a mottled maroon and yellow, while the spiny spadix is pure yellow or dark purple.

Chemicals in the flower act as a heat generator to melt snow surrounding the plant—a process known as thermogenesis. Temperatures within the flower are often 60 degrees higher than the outside air. Skunk cabbage also has a chemical called cadaverine (cadaver) that is the same substance found in decaying animal matter. The species name, *foetidus*, in Latin means putrid or to stink. This rotten meat smell attracts flies and bees that inadvertently pollinate the plant. Spiders often hide inside the warm flower to capture unsuspecting insects as they arrive.

After the plant has flowered, leaves appear that can grow up to several feet across. These huge cabbage-like leaves allow the plant, through photosynthesis, to produce more food for the root system and to store for the following spring. The skunk cabbage has an unusual root system that contracts as it grows, pulling it further into the ground and making it almost impossible to dig.

The Menominee tribe tattooed skunk cabbage dye under their skin to prevent diseases from returning, and the Ojibwe used the leaf as an underarm deodorant. For many of us, seeing this broad splash of green in early spring amid the brown and often barren wetland helps us chase away the wintertime blues.

Lady Slippers

photo: John Toren

Lady slippers, with their elegant beauty, are the rare floral gems of our native habitats. The showy lady slipper (*Cypripedium reginae*) is the Minnesota state flower. It prefers wet areas, while the yellow lady slipper (*Cypripedium parviflorum*) does well in dry or damp areas. The leaves of the showy lady slipper rise from the base of the stalk and have brown sepals, while the yellow lady slipper has purplish sepals and clasping leaves that rise up from the stalk. Hairs along the stem of both species contain a fatty acid similar to poison ivy that's toxic and keeps animals from foraging on them. Both species thrive in acidic soils.

When a lady slipper seed germinates, it must be joined to a specific fungus called rhizoctonia, which provides nutrients to the seed. The fungus eats the outside of the seed while the inner cells of the seed digest and absorb some of the nutrients that the fungus obtains from the soil. The powdery seeds are among the smallest of any flower, with one flower producing as many as 60,000 seeds. In spite of the huge amount of seeds produced, a study of 3,000 plants found that only 23 were pollinated, which helps to explain the rarity of lady slippers. In the wild it takes 10 to 17 years for a plant to bloom. While the average lifespan is only 20 years, some individual plants have lived up to 150 years.

Lady slipper roots were once regarded as an aphrodisiac and were also dried and powered for use as a toothache relief. The genus name *Cypripedium* is Greek for "Venus's shoes." An old folktale tells us that whip-poor-wills use them for shoes while walking in the woods at night. Lady slippers, also known as moccasin flowers, are now protected by law, which makes it more likely they'll be around for future generations to enjoy.

photo: *Wild Bird Journal*

Robins

I am a little robin
And my head I keep a-bobbin'
And I always rise to catch the early worm.

I've eaten all his brothers,
And half a dozen others.
Golly! How they tickle when they squirm.

– *Anonymous*

This humorous ditty matches the old refrain, "the early bird catches the worm," and of course it refers to the robin, whose arrival has become the symbol for the return of spring. His buoyant, animated song lightens our spirits after the harsh days of winter.

In many cultures it was believed that if you hurt a robin, you would get struck by lightning or your cow would get sick and give bloody milk. In Wales, if you injured a robin, it was thought that you'd be punished by witches and warlocks. In his play *Two Gentleman of Verona*, Shakespeare wrote that the mark of a man in love is that "he relishes a love song, like a robin redbreast." It was widely believed that a robin pecking at the window was not a harbinger of spring, but of disaster. We now know that a robin pecking at a window believes he's fighting a rival—actually his shadow—for territorial rights.

Most, if not all, of these folk beliefs pertain to the English robin, however—a much smaller bird than the American one, and no relation taxonomically. The two birds have two things in common: a red-orange breast and abundant distribution.

The American robin is a migratory thrush whose near relations include the bluebird, the veery, and the hermit thrush. Audubon and other early naturalists reported that settlers in America slaughtered robins by the thousands, clearly not influenced by the superstitions of their European ancestors.

Robins, like eagles and ospreys, suffered from the indiscriminate spraying of DDT. Earthworms ingested the pesticide, and robins ate the worms. The chemical was banned in 1972, and since that time robin numbers have rebounded dramatically. Today's population, estimated to be 300 million, is greater than it was in pre-colonial times.

The return of the robin to Minnesota each year has traditionally been considered a sign of spring, and it's both delightful and encouraging to spot one looking for worms on a stretch of turf that's been newly thawed by the rays of the sun. It's just as likely that the bird is listening for worms. Both sight and sound aid the robin in finding a meal.

Nowadays robins sometimes overwinter in the north if there is open water and a supply of food available nearby, especially crabapples. It's not unusual for a flock of robins to linger in the crabapple collection at the U of M Landscape Arboretum through the coldest days of the year.

The poet John Greenleaf Whittier said, "He brings cool dew in his little bill, and lets it fall on the souls of men." The melody of the robin enriches all of our lives.

The Minnesota Dwarf Trout Lily

photo: Jim Gilbert

With the warming, lengthening days of April and May, one can sit in the Grace B. Dayton Wildflower Garden at the Minnesota Landscape Arboretum, listen to the trickling water as it tumbles down the hillside, and almost hear the forest floor move as the ephemeral spring flowers emerge through the leaf litter of the previous autumn. Among the wildflowers that take part in this renaissance is the Minnesota dwarf trout lily, one of the rarest flowers on earth. It occurs naturally on fewer than 600 acres of woodland habitat in Rice, Steele, and Goodhue counties.

Research suggests that the Minnesota dwarf trout lily evolved from the larger white trout lily about 9,000 years ago following the retreat of the glaciers. It was first collected in 1871 by Mary B. Hedges, a botany teacher in Faribault, who sent a specimen to the famous botanist at Harvard, Asa Gray. Gray named it *Erythronium propullans. Propullans* is Latin for "sprouting forth," which describes how the plant reproduces. Unlike most flowering plants, the Minnesota dwarf trout lily almost never produces seeds but is spread when the underground stem of a flowering plant produces a runner bearing a new bulb.

Since only about one-tenth of the plants flower annually, reproduction levels are very small in number. It's easy to understand why the dwarf trout lily is now endangered: it lacks seeds to be spread by wind, animals, and other means.

The common name "trout lily" possibly comes from the purplish blotches on the leaves, which resemble the glistening pattern of scales on some species of trout. "Colonies" of these plants were moved to the Arboretum for research and preservation, and also to make it easier for visitors to enjoy this incredibly rare and delicate pale-pink flower.

Thawing Frogs

In the snow-melting days of late March, the calls of Canada geese and the quacking of mallards fill the air, and the wetlands and ponds around the Arboretum begin to escape winter's icy hold. One remarkable creature that rises Lazarus-like from the dead at this time of year is the western chorus frog. Chorus frogs overwinter under leaf, litter, and rocks within a hundred feet of the water they inhabit during the non-winter months. When temperatures drop below 32 degrees the frog's liver begins to produce sugars (eventually glucose) which keeps cells from dehydrating and shrinking. As the frog freezes, its heart continues to pump the protective glucose around its body. Eventually the heart stops and all other organs also stop functioning. The frog uses no oxygen and appears to be dead . . . literally a frogsicle! But as the weather warms again, the frog thaws.

photo: Bill Reynolds

A chorus frog

The call of the western chorus frog is similar to the sound of a thumb being run down the teeth of a comb. When you approach a pond of singing frogs, notice how they all stop in unison.

Frogs need fish-free waters to survive. Individual females can lay hundreds of eggs over a season (usually five to twenty at a time.) It takes eight to ten weeks for tadpoles to morph into frogs.

The clamorous trilling of the chorus frog is a springtime performance not to be missed.

The Phoebe

As snowbanks retreat under the heat of the strengthening sunshine, fields are exposed and wildflowers dot the sodden forest floor. Among the earliest of the songs we hear from returning migrants, when the first snow trilliums, hepaticas, and bloodroots are emerging and the maple sap is starting to flow, is the hoarse, exhuberant *fi-bree* call of the eastern phoebe, newly returned from its winter range in Mexico and the southern U.S.

This leaden gray flycatcher with the dingy white breast is likely to return to the same nest area every year—often under the eve of a vacation home, where every slamming screen door sends it scurrying for the woods. Other favored sites are bridges, old barns, and outbuildings with ledges—any place that offers overhead shelter. Phoebes have been nesting at the Ordway Education Building of the Arboretum for at least 30 years. Of course, the species was doing just fine before humans arrived, but the ones we notice most often are those that have chosen to live right next door. Their tightly constructed nest is made of grasses and hair and is covered with mosses. Phoebes usually nest twice in Minnesota with a clutch of four to five pearly white eggs. It is not uncommon for the parasitic cowbird to lay its egg in the phoebe's nest.

Birders are thankful that the eastern phoebe invariably pumps its tail up and down when perched. That makes it easier to differentiate it from its look-alike cousin the eastern wood pewee. By the time the wood pewee returns in mid-May, the phoebe has already established a nest. Phoebes, like other members of the flycatcher family, will fly from a perch and nab insects out of the air. This method of feeding is called "hawking."

In 1803 John James Audubon watched phoebes return to the same place on his Pennsylvania farm. He tied silver threads to the legs of several nestlings, and when they returned the next spring, the first successful bird banding was accomplished! The methods have changed, but bird-banding is a valuable tool for bird researchers even today.

When Henry David Thoreau saw an abandoned phoebe nest in wintertime, he asked, "And where is that little family now with its cradle full of snow." Not to worry, our little feathered friend will always return in spring.

Spring Warblers

If you happen to see a tiny bird flittering through the trees in your backyard in mid-May, there's a good chance it's a warbler. This extensive family of birds winters in Central America and heads north every spring at about the same time, regardless of the weather. Some warbler species nest in the Twin Cities area, but many more are intent on reaching nesting grounds farther north.

The yellow-rumped warbler is the first to arrive every spring and also among the easiest to identify: it has a yellow rump. Other warblers also carry names that can help us identify them, for example, the chestnut-sided warbler, the bay-breasted warbler, the yellow warbler, and the black-and-white warbler. But many warblers are largely yellow, and several are black and white, so it takes some study and careful observation to sort them out, and the presence of the less-flashy females further complicate matters. Quite a few warblers were first sighted in the East and carry names that have nothing to do with their appearance: the Nashville, the Connecticut, the Tennessee, and the Cape May, for example.

In short, the challenges of identifying warblers can be considerable, yet the pleasure of seeing one can be great, especially if you happen to get a good look. Figuring out which warbler it is merely adds to the fun.

photos: David Brislance

From top: black-throated green warbler, blackburnian warbler, ovenbird, northern parula

Swallows

photo: Linnette Hulbert

Feed me first! No, me!

A cherished sign of spring, the swallow is as beneficial as it is beautiful. A U.S. Agriculture Bulletin once stated: "No complaint has ever been made that these birds harm either wild or cultivated fruit or seed or that they injure other birds. It is doubtful if there is a more useful family of birds in the world."

Why? Becasue swallows eat harmful insects. They feed almost entirely on the wing, keeping their short, broadly triangular bills open wide to scoop bugs from the air. Swallows prefer houseflies and horseflies, but they will also eat plant lice, tree hoppers, ants, moths, and other harmful insects. They also drink and bathe on the wing, and it's fun to see them take a sip as they fly over water. At 10 years of age, a swallow has flown enough miles to go around the world 87 times.

Six species of swallows visit Minnesota regularly. Of these only the tree swallow consumes any appreciable vegetation. Cliff swallows are colonial nesters, building a unique gourd-shaped nest of mud, often attached under eaves of buildings or bridges. They have a tan rump patch that we might compare to the face of a cliff—a convenient memory device.

The tree swallow, with its deep-blue back and white underside, migrates to the Gulf Coast and on into Mexico. In Puerto Morelos on the Yucatan Peninsula, where I spend part of every winter, the swallows return to the mangrove swamps bordering the town at dusk each night and congregate in a flock of up to 10,000 birds. The swallows form a tornado-like funnel before they descend to their roosting site—the entire spectacle takes about a minute.

The tree swallow's nest is made of grass and white feathers whose tips curl over four to six pure white eggs. It's not uncommon for non-parenting tree swallows to help feed others' nestlings.

Barn swallows are the most abundant and widely distributed swallow species in the world. They nest in every state except Hawaii and Florida. Its luminous cobalt blue back and coppery orange breast make the barn swallow one of the beauties of the avian world. It's also the only swallow species that has a deeply forked "swallow tail."

In British Columbia, a pair of barn swallows built their nest on a train that ran two miles from Lake Tagish to Lake Atlin. The swallows would wait for the train at Lake Tagish and board it whenever it arrived. Apparently the conductor did not ask to see their tickets.

Barn swallows migrate as far as Argentina, covering up to 600 miles daily. Upon their return, they build nests of mud pellets mixed with grass, often making a thousand or more trips for construction materials before finishing the job to their satisfaction. Considering the labor involved, it's not surprising that they'll reuse their old mud-pellet nest from the previous year if possible. And if they do, they might have time to raise two broods instead of one.

In many Eastern cultures, a mixture of herbs and swallows has been thought to have supernatural healing powers. In the 17th century a treatment for epilepsy was made with a hundred swallows, an ounce of castor oil, and white wine. (At least one of the three ingredients sounds good!) German peasants believed that if a swallow flew under a cow, the cow would sicken and die.

Dreaming of swallows means you'll have family happiness. Perhaps instead of counting sheep we should count swallows as they dip and dive through our ethereal world of slumber.

photo: Tom Sanders

A barn swallow feeding its young on the wing

It's Good to Be Queen!

photo: Zachary Huang

The exuberant calls of robins and red-winged blackbirds are familiar harbingers of spring. A quieter sound that we often overlook is the humming of honeybees as they gather nectar from pussy willow blooms. The buzzing and bouncing of these insects on willow catkins is an overture to Mother Nature's spring symphony.

Honeybees are not native to the U.S. They were first brought to America in 1638 by John Smith, who wanted to provide "sweets" for the Virginia colonists.

Here are some fun honeybee facts: It takes 40,000 bee loads of nectar to produce one pound of honey. A productive colony can add 70 pounds of honey a week to their hive. A pound of clover honey represents the distilled nectar of 8.7 million flowers. A queen bee can lay up to 3,000 eggs in a day. When the queen lays an egg she can determine the sex and kill any possible rivals. Queen bees can sting many times, while sterile worker females sting only once. Males (drones) can't sting at all and are routinely killed after mating. The average life span of a worker bee is three months. The almond industry uses 1.4 million colonies of bees—approximately 60 percent of managed hives in the U.S.

Colony collapse disorder has had a devastating effect on bees, and also on the plants they pollinate. The new Tashjian Bee & Pollinator Discovery Center at the Arboretum, which opened in 2016, will be a leader in determining the causes of bee decline and devising methods to reverse the trend.

Aspens

photo: John Toren

Aspens in spring on a hillside at Frontenac State Park

Aspens are the most widespread and abundant tree in North America, ranging from the edge of the tundra to northern Mexico. They make up about a quarter of the northern forest. As a sapling the quaking aspen's bark is pearl-gray and white, becoming knobby and black at the base with age. In late winter the aspen produces fuzzy ash-gray buds that in springtime burst into leaves, vivid green above and dull green below.

Deer and moose browse on the twigs, snowshoe hare nibble the bark, and beavers strip the bark for food before toppling the tree itself to use in constructing their lodges. Every 10 to 16 years, forest tent caterpillars strip the leaves, denuding acres of trees. The barren branches become entangled in the silken nests of the bluish caterpillars. This devastation can be seen most notably in the forests of northern Minnesota, where the tree is commonly known as a popple.

Aspen bark, like the bark of other members of the willow family, contains salicin, the natural ingredient in aspirin. Many Native American tribes traditionally used it to soothe coughs and colds and also made a tea to treat urinary problems. In the 17th century a herbal remedy book, *The Doctrine of Signatures*, stated that parts of the body could be cured by plants that resembled them. The quaking aspen was therefore used to treat ague and other diseases of constant shaking.

The long, flattened leaf stalks of this tree, which are unique, account for this picturesque dance. When the golden-yellow autumn leaves of the quaking aspen are bathed in sunlight they resemble miniature ballerinas pirouetting in diamond-studded tutus.

Being a soft wood, aspen's commercial uses are limited to the pulpwood and fiberboard industry. This sector has been increasing in importance in recent decades, as mature forests become more scarce. Aspen also has the merit of regenerating itself from stumps and roots. Many stands of aspen have a massive cloned root system. In Minnesota, one system has been determined to be roughly 8,000 years old. However, the age of this vast organism pales in comparison to a quaking aspen system in south central Utah called Pando. (In Latin *Pando* means "I spread.") At 6,600 tons, it is considered to be the largest and heaviest organism on Earth with the possible exception of fungal mats in Oregon. It covers 106 acres, has more than 40,000 trunks, and is at least 80,000 years old, though some researchers speculate that it may be nearly a million years of age. In honor of this gigantic specimen the U.S. Postal Service issued a stamp in 2006 as part of the "Wonders of America" series listing it as the largest plant in the country.

photo: NW nativeplant database

The Chipmunk

photo: David Brislance

In many ways, the real-life eastern chipmunk resembles its cartoon counterparts Chip and Dale. Chipmunks hold their tails in the air as they scurry about filling their saddlebag-like cheek pouches with a variety of seeds, dried fruit, and nuts. Their pouches can stretch to three times the size of their heads. One chipmunk was reported to have 70 sunflower seeds in its cheeks. They store these packets of food for winter use. Tamias, the genus name of the chipmunk, means "storer."

Chipmunks spend most of their life on the ground. They build burrows, usually three feet deep and 10 to 40 feet long, and cover the entrance holes with leaf litter to hide them from predators. The dens are divided into separate chambers: bedroom, kitchen, and bathroom. Small remnants of acorn shells or seed husks on a stump or rock are a sign that a chipmunk has found a tasty morsel to munch on. Chipmunks are friendly little critters who can be bribed with food and tamed somewhat.

Chipmunks are light hibernators, unlike their squirrel cousin the woodchuck, who goes into deep sleep for almost six months every winter. Chipmunks will awaken from their torpor every two or three weeks to feed and then return to bed. They're solitary animals, and the "chipping" noises you hear them make often come from individuals defending their territory. In Minnesota the female will typically have one litter of two to five offspring sometime between February and June. When the young are born they're roughly the size of bumblebees. Once they reach maturity their mother drives them off to find territories of their own.

A Native American legend tells of the time a chipmunk made fun of a bear by betting him he couldn't keep the sun from rising. The bear stayed up through the night but at dawn the sun came up and the chipmunk laughed at the bear's stupidity. In his anger, the bear took a swipe at the chipmunk with his paw, which explains how the chipmunk got three stripes down its back.

photo: Carol Knabe

Ospreys

Ospreys are among the most exotic and appealing of the raptors, in part because of their facial mask, in part because of their dramatic aquatic hunting techniques, and also because they have no close relatives—they're the only living member of their family.

Ospreys feed almost entirely on fish. Their feet are very large and strong, and the soles are armed with sharp spines. They plunge feet first into the water, dropping from heights of up to 100 feet to seize unsuspecting fish, occasionally becoming entirely submerged before they've secured their "catch."

Ospreys build nests in treetops or at other high points with unobstructed views. Bald eagles, by contrast, build their nests lower in the crotches of trees—one way to differentiate their nests. Ospreys can live up to 40 years and may use the same nest throughout their lives.

Enemies of the osprey include the bald eagle, which will consistently steal food from the smaller bird and generally harass its nesting site. Great horned owls will kill osprey at night. Osprey populations were severely impacted by human use of DDT and other pesticides. Fortunately, their numbers have increased since DDT was banned in 1972.

Ospreys migrate as far as South America and return to Minnesota in late April. They lay one to four eggs, which are usually buff-white and heavily marked with chocolate. Young opsreys remain in their winter territory until their third year.

In April 2001, a pair of ospreys built a nest on an electric pole at the Horticultural Research Center of the Arboretum. With the help of Hennepin County Parks, we installed a nest site near Lake Tamarack. The ospreys moved to the new nest within hours. The female had been raised in St. Paul and was two years old; the male, five years old, came from Steiger Lake. The female was too young to lay eggs, but the pair defended the nest all summer. They returned to the nest site the next spring and fledged their first baby. The nest has been active every year since.

photo: Carol Knabe

The author explains a simple principle: When water is removed from maple sap, the result is syrup.

Maple Syruping

The time for tapping maple trees is dictated less by astronomical events per se than by the rise and fall of temperatures during early spring. When daytime temperatures start to rise above freezing (32 degrees Fahrenheit) while nighttime temperatures remain below that mark, the sap will start to flow. For the next few weeks a few ambitious folks will be out in the sugar bush, tapping the trees. When daytime temperatures rise above 40 degrees (if nighttime temperatures remain below freezing) sap flow reaches its peak.

At the Arboretum, 300 trees are tapped—200 by the sap-collecting crew using connecting lines of blue plastic tubing, and 100 by children in the youth education program using blue plastic bags. The plastic tubing is hooked up to a tap in each tree, and the sap flows by gravity downhill from tree to tree into collecting tanks. This network of tubing eliminates the need to climb the hills

photo: Mark MacLennan

and empty individual bags. To put less stress on the trees, only one tap per tree is now recommended, and the tap hole is made no more than two inches deep and at least three feet from the ground.

The average sugar maple tree that is tapped produces about 10 to 15 gallons of sap per season, but some trees can produce up to 80 gallons. The average sugar content of maple trees ranges from about 2 to 3 percent, which means that even from a good tree, it takes more than 30 gallons of sap to produce a gallon of syrup.

The process of making maple syrup hasn't changed much since Native Americans first discovered how to produce it. Water is boiled out of the sap until the temperature rises to 219 degrees. At this point the sugar (sucrose) content of the sap reaches 67 percent and we call it maple syrup.

No one knows why maple syrup turns brown or caramel color. Researchers believe that at least one of the six sugars and one or more of the 12 organic acids in maple sap may be involved, but they have yet to completely solve this mystery of Mother Nature.

After tapping, a maple tree doesn't actually "heal" the wound. Instead, new growth occurring under the bark seals off the hole so it won't become an avenue for insects or diseases to enter the tree. If tapping is done correctly, a tree can be tapped for decades with no ill effect.

Those who spend time attending to this activity may notice that other natural processes have also gotten underway. The dainty snowdrops, with anti-freeze in their veins, are already in bloom, though patches of snow still cling stubbornly to the forest floor. Sharp-lobed hepaticas, bloodroot, and pasque flowers will soon make their appearance, attracting bumblebees, who keep warm in their black and yellow coats as they search for nectar.

And red-bellied woodpeckers are tapping out their love notes in Morse code fashion, hoping that a future mate is listening nearby. As the sun settles over the maple trees in the west, the robins continue to sing their cheery song until every ray of light is squeezed from the sky. As the sun sets a full moon rises in the east. Native Americans call this the "Sugar Moon"—no explanation necessary. In the sugarhouse the steam rises from the cooker. The cool darkness of evening closes around us, and as the syrup boils, we move our chairs a little closer to the fire.

Turkey Vulture

As the mild springtime breezes usher in the verdant new season, they also signal the return of many winged friends. Among the larger returnees is the turkey vulture, which use warming thermal updrafts to wend their way back to their summer breeding grounds. In migration vultures can cover 3,000 miles in 10 days without eating. They have a five- to six-foot wingspan and hold their wings in a shallow V-shaped pattern as they make a veering tipsy transit across the sky, unlike the similarly sized bald eagle, who exhibits a flat wing profile and a more steadfast soaring path. The silvery underlinings of the vulture's wings also aid in identifying them from below. But close up, turkey vultures have red, warty, featherless heads that only a mother could love.

A soaring turkey vulture can be mistaken for a bald eagle. The upturned wings give it away.

Vultures are the great recyclers of the bird world, feeding for the most part on fresh carrion. Their bodies are uniquely adapted to be nature's garbage disposals. Their bald head is an evolutionary adaptation allowing them to feed on the inside of carcasses without their feathers getting in the way. Their powerful digestive systems kill several deadly disease organisms

photo: Carol Knabe

Turkey vultures perform a useful function when they feed on roadkill.

including salmonella and bacteria that cause hog cholera, anthrax, and botulism. Unlike most birds, turkey vultures' olfactory nerves are very sensitive to smell, and they find their meals more often by smell than sight. They will not eat putrefied meat, but an animal must be dead for at least 12 hours before vultures can smell it.

Turkey vultures typically mate for life and return to the same nest site year after year, usually laying one to two eggs, often in the trunk of a dead tree. On our family farm in Wabasha, we were fortunate to have turkey vultures nest in the same tree for 12 years running. We were able to watch the life cycle of the young from egg laying to fledging. The odors emanating from the nest made it clear their apartment needed a good spring cleaning.

Vulture legs usually look white because they defecate on them to regulate body temperatures. When threatened, vultures will vomit on themselves and sometimes play dead. Yuck!

There once was a turkey vulture carrying two dead animals onto a plane. The flight attendant stopped him and said, "Sorry, sir, only one carrion is allowed per passenger."

photo: Carol Knabe

Canada Geese

Ornithologists have distinguished seven subspecies of Canada goose, the largest of which, the giant Canada goose, is a common resident and migrant in Minnesota and Wisconsin. The bird was at one time on the verge of extinction, and in 1932 Thomas Roberts wrote in *The Birds of Minnesota*, "With a single exception there are no reliable nesting records... their numbers being so greatly reduced that a migrating flock of Geese is now an event of special interest over much of the state."

Times have changed. Repopulation efforts were started at Lake of the Isles in Minneapolis in the early 1970s, and now approximately 25,000 giant Canada geese live in the Twin Cities metropolitan area. Without controls, this population would swell to 100,000 or more. Since the diet of the Canada goose consists mainly of grass, the birds have flourished as the number of lawns, golf courses, and parks has increased. The rise in goose populations has also increased the danger of goose collisions with airplanes.

Canada geese mate for life and produce one brood a year of two to seven eggs. The female incubates while the nearby gander defends her from intruders. Geese mate in their third year. Non-nesting geese seen in spring and summer are two-year-old bachelors and bachelorettes. The average life span is 10 to 24 years.

Goslings communicate with their parents while still in the egg. They have several calls including a distress call to alert the mother when the egg gets too cool. A distress call is also used when goslings have difficulty getting out of the shell or, once out, have trouble keeping up with the parents. The goslings make a trilling comfort call when they're going to sleep.

Goslings also communicate with one another within the egg to synchronize hatching. By this stage in their development they have an enamel egg tooth that they use to work their way out of the shell.

New goslings remain in the nest a day or two to remove a thin sheath that covers their down feathers, making them look wet when they emerge from the egg. By cuddling against the mother and moving around in the nest, they rub off this sheath, leaving fluffy little balls of gosling.

Imprinting, a learned, irreversible process of recognition, was first studied in geese. It takes both goslings and parents about two days before imprinting occurs. Geese will accept foster goslings during this time, but will bite them and chase them away once imprinting has occurred. Sometimes geese will form crèches—a number of goslings with a few parent pairs.

Geese fly in the V formation, not because it's the only letter of the alphabet they know, but because the lead bird creates turbulence that allows the following birds to expend less energy. The V formation also makes it easy for members of the flock to maintain visual contact.

Geese and their eggs have been eaten and their fat used for cooking oil for millennia. Goose feathers were once widely used for quill pens, and their soft down is still a prime material for insulating outdoor clothing. The Romans used geese as sentries. When raiders tried to enter the city in 387 BCE, the geese raised the alarm.

Geese have also starred in many a myth and legend. We all remember Aesop's fable about the man who killed the goose that laid the golden eggs. In Egyptian mythology, a goose laid a golden egg that became the sun. Hindus believe the Great Spirit laid a golden egg from which the god Brahma was born. Mother Goose, whom we associate with all sorts of tales and rhymes, recalls Bertha Goosefoot, the mother of Charlemagne, who allegedly had big floppy feet and was called Mother Goose by the peasantry—after her death, of course.

photos: John Toren

A Walk along the Zumbro River

The Zumbro River is an unpredictable waterway. It can be transformed by spring rains from a gurgling brook with log jams, sandbars, and gentle eddies to a tempestuous roiling flood. When the river's fury finally dissipates and it returns to the confines of its established banks, signs of spring abound.

If we venture that way, we're likely to see thousands of dainty spring beauties covering the forest floor with a pinkish hue. Swamp buttercups, with their rich yellow blossoms, gleam in the dappled sunlight. Bloodroots are sprinkled among the patches of false rue anemone. Known as "the wind flower," the anemone contains no nectar but regenerates itself by the breezes that spread its pollen. A Greek legend tells us that Zephyr, the god of the wind, fell in love with the nymph Anemone. This made the goddess Flora jealous, and she turned Anemone into a flower. We might also spot a bed of white trout lilies or a lone Virginia bluebell.

Other flowers of the river bottom environment attract bumblebees and red admiral butterflies searching for their own nectar treats buried in the blossoms. No

photos: John Toren

(left) A swamp buttercup; (right) Virginia bluebells

one could miss the evidence of beaver activity; a family has gnawed its way through a stand of basswood trees. Some teeter on the brink of falling, while others have fallen and been dragged away to be consumed or to add bulk to a nearby dam or beaver house, leaving only stumps behind.

In places where the sun shines across the river we see painted turtles warming themselves on stones protruding from the bank or on branches of overhanging debris. Birdlife has also returned. During our walk we disturb a belted kingfisher, who clatters loudly as it flies to a quieter place along the river. Yellow-rumped warblers—usually the first warbler to arrive—have returned from their winter in Central America to feed hungrily on the insects attracted to the budding trees. And on the other end of the bird spectrum, we see a bald eagle that has spent the winter down on the Mississippi as it surveys the inland waters of the Zumbro for an unsuspecting duck or fish.

photo: David Brislance

A yellow-rumped warbler

Summer

photo: John Toren

Big bluestem and other prairie plants flourish at Touch the Sky National Wildlife Refuge north of Luverne, Minnesota.

Summer Prairie Walk

In springtime, when prairies are covered with the dead plant material of the previous year, resource managers often schedule controlled burns to help generate new growth in the same way that lightning strikes and the hunting techniques of Native Americans did prior to white settlement. It's difficult to imagine how the black, sooty soil of a spring burn can turn so quickly into the spectacular swaying grasses and colorful flowers of summertime—but it does. Many prairie species are deep-rooted perennials with their growing point at or beneath the surface, allowing them to survive fire, wind, hail, and grazing. As much as 85 percent of the biomass in a prairie is underground. For example, little bluestem and switchgrass root systems can grow to a depth of five feet, while leadplant and compass plant can penetrate ten feet into the soil.

The tallest and most impressive of the prairie grasses is big bluestem. It's also called "turkey foot" because of its three-pronged seed head, which resembles a gobbler's claw. When pushed by a breeze, the undulating waves of big bluestem make the prairie come alive.

The silphiums (a Greek word whose meaning has been lost) are the goliaths of the prairie flowers, towering above other species. The family includes compass plant, prairie dock, and cup plant. Their yellow daisy-like flowers on stout stalks can grow to eight feet. Leaves of silphiums are bristly like sandpaper, an easily identifiable feature. Cup plant leaves are joined at the base, forming a cup holding water where goldfinches, chickadees, and other birds drink. Some Native Americans used a blob of the resinous sap as chewing gum or to freshen breath.

(above) During a rain, water collects at the base of the cup flower's leaves, which wrap completely around the stem. (below) A dewy leadplant catches the morning sun.

photo: John Toren

Another plant that grows in large groups is rattlesnake master. The leaves of this plant are spiny-edged, and the flowerheads resemble the armor-piercing mace of medieval knights. Rattlesnake master got its name from the Native American belief that it could bring relief from snake bites.

The prairie is also home to flowers with blue, purple, and pinkish hues including purple coneflower, leadplant, blazing star, and purple prairie clover. Purple coneflower was called "thirst plant" because when its salty root was chewed it increased the flow of saliva, alleviating thirst.

The thick root of lead plant was so difficult to plow that the pioneers called it "devil's shoestrings." Purple prairie clover is native to Minnesota, though most local clovers were introduced and have become naturalized.

Only about 1 percent of our pre-pioneer prairie remains undisturbed, but efforts to reintroduce native prairie grasses have been widespread and largely successful. Prairie "corridors" to strengthen bee populations are being developed in western Minnesota, and many regional and state parks now have prairie "patches" where you can take a stress-reducing walk through the open countryside, observing a variety of unusual plants, birds, and insects that call the prairie home.

Dragonflies

While walking through the prairies and wetlands on a summer day, it's fun to observe the dragonflies. They can propel themselves in virtually any direction, including backwards and sideways, and resemble tiny helicopters as they dash and dart about. They are in the order Odanata, which means "toothed ones." They snare most of their food on the wing, catching small insects with their mouths and using their hairy spiked legs as a net to snag larger insects. Dragonflies might well be described as "mosquito hawks," being major predators of that pesky insect.

The eyes of the dragonfly are composed of nearly 30,000 lens, making it possible for them to scan 360 degrees laterally and also up and down. They can see in ultraviolet light, polarized light, and dim light. Since they can't hear, smell, or vocalize, these large eyes are their only means of communication. Dragonflies drink by dropping onto the surface of a body of water and then absorbing moisture through their exoskeleton. When mating, the male grips the female by the head while their abdomens meet and her eggs are fertilized. This process can take as little as 10 seconds. In most species the female will then dip her abdomen into the water to wash off the eggs.

Most dragonfly nymphs mature to adulthood in one to three years and molt numerous times during their development. Like the adults, they are aggressive hunters, eating mosquito larvae, tadpoles, and even small fish. Dragonflies do not overwinter as adults in Minnesota; most species survive in larval form in suspended animation beneath the ice. The large green darner is one species that does

photo: David Brislance

A 12-spotted skimmer alights on a lupine flower.

migrate to the south with its offspring returning the next year. A green darner migration can number in the thousands, and if you've been lucky enough to see one, you know that it's a spectacular sight. In Minnesota green darners and nighthawks migrate at the same time, both feeding on smaller insects as they head south.

Fossil records of dragonflies go back at least 300 million years. One species, with a wingspan of 30 inches, is considered to be the largest insect ever to have lived. As children we were told dragonflies would sew our ears shut if we didn't behave. In Swedish folklore dragonflies are said to pick out human eyes. Norwegians call them "eye pokers." But not all societies consider them fearsome. The Navajo consider them as symbols of pure water. Indonesians eat them as a delicacy. Legend has it that a former emperor in Japanese named Japan "Akitsushima" —the "Isles of the Dragonfly."

Dragonflies are harmless to humans, so don't be alarmed if one lands on your leg or shoulder. Rather, take the opportunity to examine closely this fascinating creature that feasts on mosquitoes and other little critters that make our summers less enjoyable.

The Pocket Gopher

Minnesota's plains pocket gopher has been considered an agricultural pest for decades. Since 1864 more than a hundred types of traps have been developed to eliminate it. Only gradually has it become widely recognized that while the pocket gopher can do damage to some crops, it also has many beneficial effects on the environment.

The pocket gopher is a burrowing rodent that lives most of its life underground. It has been called "the aerator of the plains," as its extensive digging is a key component in the health of any ecological system where it is present. The exposed soil increases plant regeneration and distribution and allows rainfall and snowmelt to more efficiently permeate the subsoil.

The gopher uses its head as a miniature bulldozer to push the soil to the surface. One gopher can dig up to 50 mounds. A study in Yellowstone National Park estimated that one pocket gopher can excavate as much as five tons of soil each year. Think of the time it would take to remove an equivalent amount of soil with just two teaspoons! A pocket gopher's tunnels can total up to 500 feet in length. These runways are usually 4 to 12 inches below ground, though the storage area can be as deep as 6 feet. The storage area has three rooms—a pantry, a toilet, and a sleeping space—just like a miniature efficiency apartment. A study in Colorado documented at least 22 species of animals utilizing abandoned gopher burrows, including snakes, toads, mice, and ground squirrels.

The pocket gopher's body is uniquely adapted to an underground existence. It has powerful front legs with sharp claws and lips that close behind its sharp incisors, keeping dirt out while it digs. It has a tube-shaped body, narrow hips and smooth fur that allow it to easily move back and forth in cramped spaces. The "pocket" is actually a fur-lined cheek pouch used to store food. Pocket gophers are solitary animals, with females having one to six young annually.

Because pocket gopher mounds caused havoc in hayfields, farm kids used to trap gophers and receive a bounty for each pair of front feet taken. Some Minnesota counties still offer a bounty—usually about enough to buy a pack of gum.

photo: National Park Service

A pocket gopher's mound is horseshoe-shaped. A vole's mound is symmetrical.

photo: Carol Knabe

Bobolink

The male bobolink appears to be dressed like a black and white police car with a dollop of buttery yellow on the nape of his neck. As striking as his physical appearance may be, it is his song that makes him the Caruso of the fields and meadows. At the turn of the twentieth century, New Hampshire naturalist Schuyler Mathews proclaimed it "a mad reckless song-fantasia, an outbreak of irrepressible glee," and ornithologist Alexander Skutch declared that "his notes seem to stream out so rapidly that they fall over each other and become jumbled."

Although complex bird songs are difficult to capture in words, these poetic phrases are a fitting tribute to this meadow maestro. Sadly, because of habitat destruction in agricultural areas, the bobolink's song is heard less and less these days. Breeding bobolink numbers have dropped about 75 percent in the last 25 years. A study in New York found that 80 percent of bobolink nestlings survived in undisturbed fields compared with only 6 percent in fields that had been mowed and baled.

The genus name for the bobolink is *Oryzivorous*, which means "to eat rice." In the early 1900s bobolinks would feed on the milk stage of the rice crop in the South; millions were killed by irate farmers who shot at them from dawn to dusk. In Jamaica they were called butter birds and killed to be eaten, even though a bird could provide only a single mouthful of food.

The migration flight of the bobolink totals at least 11,000 miles, the longest of any North American songbird. Bobolinks have magnetite in their nasal tissues that can act as tiny compasses. Researchers are still trying to figure out how the birds make use of the magnetic information.

This joyous little songster, which often flutters in the wind above the fields as it sings, is on every birders must-see list—a well-earned accolade, for sure.

A Summer Walk on the North Star Trail

I know a place where two trails meet. From this viewpoint, in the coolness of a mid-August morning I can see a wetland surrounded by forested hills that protect its natural quietude. Dew-covered spiderwebs stretch across it, glistening in the early morning sun like a sheet of diamond-studded gossamer. Its border is edged in the rose-purple flowers and spiny leaves of bull and swamp thistles that act as sentries against intruders, intermixed with stinging nettles. Anyone who has felt the fury of this fiery plant will understand the origin of its Latin genus, *Urtica*, which means "it burns."

The "danger" associated with these plants is largely an illusion, however, and the wetland itself can be a soothing and welcoming place to visit. Here the tiny male yellowthroat, with his black Lone Ranger mask, flits from stem to stalk while greeting the new day with his energetic whitchity-whichity-whichity call. Here the merry song sparrow quickly appears and disappears in and out of the undercover, challenging the best of bird watchers to correctly identify him. Ruby meadowhawk dragonflies and male widow skimmers, with their black and white wings, sit motionless on the tips of sedges, warming themselves in the sunbeams before beginning their daily forays against the pesky mosquitoes. They are truly a natural insecticide.

In mid-August the milkweed is in bloom, and if we look closely under the leaves, we're likely to find the pale green lemon-shaped eggs of the monarch butterfly delicately attached there.

As we walk along the trail, we hear what sounds like the buzzing of a telephone line. In reality it's the annual cicada beginning its daytime serenade. Fresh raccoon and white-tailed deer tracks are imprinted in the soft parts of the path, letting us know we're not alone. Our movement along the trail is marked by a measured silence, as if the creatures of the forests are saying "Man walks among us. Be still! Be still!"

photo: Carol Knabe

A common yellowthroat

If we look farther into the wetland, Joe Pye weed stands out as the colossus amongst the flora, sometimes reaching six feet in height, its tight cluster of pinkish-purple blooms tower over lesser plants. "Jopi" is a Native word for typhoid, and Joe Pye appears in folk tales as a Native American who used the plant to treat the disease.

At the end of the trail there is a picnic bench surrounded by a stand of majestic white pines—a place to take a short rest, forget about our problems, and enjoy being alive in nature. The pines sigh as a light breeze caresses their soft needles, and a nervous chipmunk scurries onto the bench looking for a handout, as if the bench were his and we must pay him for the use of it. It's peaceful here in Mother Nature's secret garden.

photos: John Toren

(left) Milkweed; (right) Joe Pye weed

Cicadas

photo: Carol Knabe

The high-pitched whining drone of the annual cicada resembles the buzz of a power line, though it tends to be more intense. It starts softly but grows louder over a span of time to reach an ear-piercing crescendo, then tapers off more abruptly. This army-green and chocolate-brown insect, an inch or two in length, is also called the dog day cicada because its peak singing activity is during the muggy days of August, when Sirius, the Dog Star, in the constellation Canis Major, reaches its highest point in the night sky.

Unlike most crickets, who generate sound by rubbing their wings together, the cicada (whose name is derived from the Latin for "tree cricket") uses sound organs called tymbals to produce the humming sound that attracts females. The mating period is the final stage in a most unusual life cycle. Once mating has occurred, females insert their eggs into small tree branches that die and drop to the ground. When the nymphs hatch they tunnel into the ground, where they may stay for up to eight years, feeding on tree roots. They undergo several metamorphoses during this period before reaching adulthood.

In the final stage of development, the cicada constructs an exit tunnel to the surface and emerges. At that point it sheds its skin once more—the crusty brown shell of the last metamorphosis can sometimes be found still attached to a tree trunk.

In China these shed skins were traditionally used to treat ringing in the ears. Babies were also given a cicada tea to keep them quiet as quiet as cicada nymphs. Mayans and Aztecs placed jade cicadas on the tongue of the deceased, believing that such a display would insure new life to their loved one, just like a cicada nymph coming out of the ground.

photo: David Brislance

The Industrious Beaver

In the depths of winter, you might see a six-foot-high mound of branches, rocks, and mud protruding from the bleak landscape of a frozen lake or pond—a beaver lodge. Beavers spend much of the winter in these dens, the inner chambers of which are above water level, though accessible only by means of tunnels leading into them from beneath the water and ice.

The resident beavers have spent the summer consuming aquatic plants and felling trees in order to gain access to the tender leaves and shoots high above them. They also cache branches underwater as winter approaches to help them survive during the snowy months.

Beavers have chisel-like incisors similar to other rodents that continue to grow throughout their lives. However, unlike the beavers depicted in cartoons that chew through a log in a few seconds, it can take a real beaver two or three nights of work to gnaw down a large tree. They have no idea where it will fall, and some have been found crushed under a tree they've toppled.

The beaver uses its paddle-like tail to warn others of danger and also as a rudder while swimming and a brace while cutting down trees. The tail contains a large amount of fat, making it an emergency food shelf.

A beaver dam is an engineering marvel that not only provides a home for the beavers, but also raises the water level, creating ponds and expanding the sources of accessible vegetation.

The largest beaver dam ever found was at Three Forks, Montana. It measured 2,140 feet in length and was 14 feet in height and 23 feet thick at the base. Radiocarbon dating of wood fragments from a dam in California suggests that it was first built around the year 580, was repaired in 1730, and was abandoned after 1850, more than 1,200 years after it was built.

Beavers have adapted well to an aquatic lifestyle; their fur coat has a dense inner layer that water cannot penetrate, and beavers also groom their coats with castorum, an oil secreted from anal glands that further waterproofs their fur. Castorum contains salicylic acid, the active ingredient in aspirin, and was used as a headache remedy in the 1700s. Castorum oil is still widely used in the manufacture of perfumes.

Beaver pelts were valuable commodities during the 18th and early 19th centuries, when hats were fashionable. Extensive trapping almost obliterated the once abundant species, though it also spurred the European discovery and settlement of many parts of North America.

photo: John Toren

Cardinal Flower

Roger Tory Peterson called it "America's favorite," and in the 1940s, American botanists voted it the most beautiful flower in the world. When the brilliant scarlet cardinal flower blooms, it's easy to understand the fascination the experts had with this preeminent beauty. The dazzling red flowers grow in a spiked cluster, each about an inch across, with three spreading lower petals and two upright ones. The petals join in a basal tube.

Hummingbirds are attracted to red colors and are the chief pollinators of the cardinal flower. Its deep nectar-filled tube is perfectly suited to the hummingbird's three-inch bill. The red color also makes the cardinal flower unappealing to many insects. (Insects are attracted to flowers that strongly absorb ultraviolet light. The color red does not.) As an added bonus, the leaves of the cardinal flower secrete a toxic white latex that makes it unpalatable to animals.

Native to North America, the cardinal flower was introduced into Europe during the 1620s and named after the bright robes of Roman Catholic prelates. Native Americans traditionally ground the plant into a powder and used it to dispel storms. A tea was made from its roots to treat epilepsy and typhoid and to expel worms. The root was also used as an aphrodisiac, especially by elderly women. The plant's chemical properties are now being studied as a possible treatment for neurological disorders.

The cardinal flower has no scent, but you can look for its showy radiance in July when it first blooms along stream banks and wetland meadows.

photo: Carol Knabe

The Goldfinch

A common sight and sound in our Minnesota skies is the goldfinch. This little lemony ball of abundant energy, with its inky black wings, sings a *swee, swee* song while it dips and dives through the air, as if it was riding on a roller coaster. Prior to 1960, when it was replaced by the common loon, the goldfinch was the official Minnesota state bird. Regardless of its official rank, the goldfinch remains one of our most beneficial species, due to its diet, which includes thistle, dandelion, and ragweed seeds, along with cankerworms, tree aphids, and other harmful insects.

The goldfinch nests later than any other native Minnesota bird species. Nests can still be active into September. At the Arboretum we have found young nestlings the first week in October. Its timing may seem off, but the goldfinch raises its brood in coordination with the seed formation of the thistle. Some of its nest material is thistle down, and thistle seeds make up a part of the diet of the young. The former genus name for the goldfinch was *Carduelis,* which derives from the Latin word for thistle. Its nest of bark, spiderweb, cattail, thistle and milkweed down is constructed so tightly that it holds water, and the nestlings could drown in a downpour if the nest weren't covered by the parent.

After leaving the nest, young goldfinches are fed by the father for two or three more weeks. The continual *chiporee chiporee* calls we hear in late-summer are the fledglings begging for food. The goldfinch is one of the few passerine (perching) birds to undergo a pre-mating plumage molt. Beginning in late winter into spring, the male turns from a muted olive color to a brilliant yellow and black. Goldfinches join the chorus of spring birds in May even though they mate later in the year, and they've been given a nickname as a result of this trait: "warbler wannabees."

photo: Radim Schreiber - www.fireflyexperience.org

Fireflies

As youngsters, we filled jars with "lightning bugs" and watched with fascination as they flashed their little beacons. In poetic phrase, this bug is sometimes mistaken for the "will-o-the-wisp," the elusive spook that haunts the forests at night. As adults, we refer to it more prosaically as the firefly.

Male fireflies provide the aerial light show, while females flicker from the ground during the mating ritual. A clap of thunder can cause a field of fireflies to flash simultaneously, which explains, perhaps, why they're also called "lightning bugs."

The eggs and larvae of some firefly species can glow and are known as "glow worms."

The light given off by fireflies is bioluminescence. Scientists have found many important uses for the chemicals that create the light, luciferin and luciferase: among other uses, they can be used to detect harmful bacteria in food and water. They're also being tested as treatment for human diseases such as cancer, multiple sclerosis, and heart disease. Firefly technology has been used to produce safer cold light for flashlights and holiday lights.

photo: Radim Schreiber - www.fireflyexperience.org

The summertime firefly show can be seen nightly in open fields and meadows. The performances are free, courtesy of Mother Nature. We see these shows less often than we used to, however. No one knows quite why, though the decline is probably related to expanding suburbs and growing light pollution. Fireflies prefer warm, humid terrain near standing water, and as real estate is drained to make way for housing, such environments become less common.

One good place to see fireflies is at the Minnesota Lansdscape Arboretum. They put on a magnificent display in the Spring Peeper Meadow every summer.

Chimney Swifts

Unlike the other passerines (perching birds), chimney swifts are able to hang only vertically, so they're constantly in flight during daylight hours, feeding and drinking on the wing. They eat nearly a third of their body weight in flying insect pests such as mosquitoes, biting flies, and termites every day.

Swifts winter in the Amazon basin of Peru and return to the temperate climes in May. Nests are made of twigs glued with saliva. Both parents construct the nest, and both incubate the three to five white eggs. Those fortunate enough to have seen these chittering little "flying cigars" dive into and pop out of chimneys know what a wonderful experience it is.

Before humans encroached heavily on the land, chimney swifts nested in hollow trees. As the forests were logged and depleted, the birds began to nest in the brick chimneys of pioneer homes. Nowadays many chimneys are capped, however, eliminating nesting and roosting sites for the swifts. As a result, their population has declined by 50 percent in the last 40 years.

Duckweed

photo:geograpg.org.uk

Many people believe the lime-green duckweed mats that cover the tranquil waters of lakes and ponds are some form of algae that signifies a polluted aquatic system. In fact, duckweeds absorb large amounts of nitrogen, phosphates, and other pollutants, thus improving water quality.

It's also a primary food source for ducks and other waterfowl and provides cover for aquatic species like tadpoles and turtles. Painted turtles basking on a log camouflaged in duckweed appear to have just emerged from the primeval slime. The roots of duckweed are sticky, and it is readily carried to new aquatic areas on the feet and feathers of waterfowl.

Duckweeds are the smallest flowering plant on earth. They reproduce at a prodigious rate, doubling their numbers in less than two days. As the cold season approaches, the plants form buds that sink to the bottom and lie dormant through winter. Gases released during the spring thaw carry them back to the surface of the water, where they generate new growth.

Dried duckweed contains about 40 percent protein and 5 percent fat and is widely used as feed for fish and poultry. As a bio-fuel it has five to six times as much starch as corn, needs 80 percent less space than corn to grow, and doesn't contribute to global warming. It is already being used in waste treatment plants and is considered to have great untapped value for pollution control. It is also being studied around the world as a source of clean energy.

Some Southeast Asians consume duckweed, calling it Khai-nam, which means "eggs of the water."

This tiniest of flowering plants has the potential to improve the quality of life for many species on earth.

Touch-Me-Not

The touch-me-not is a succulent plant with bright yellow or orange flowers. It has a hollow stem, thrives in moist soils, and can grow up to five feet tall.

The unusual name may sound threatening, but it refers to the fact that when the plant's seedpods are gently squeezed they explode like miniature jack-in-the-boxes. The seeds can shoot out four to five feet when disturbed by an unsuspecting human, animal, or a gust of wind.

The raw juice and crushed stems of the touch-me-not relieve the irritation of poison ivy and stinging nettle—two noxious plants that tend to grow where touch-me not is abundant. The Ojibwe traditionally make a medicine from the plant for use to cure mouth sores. The seeds are edible and taste like butternuts.

Another name for the touch-me-not is jewelweed, because sparkling beads of dew often glisten on the leaves like tiny diamonds in the morning sunshine.

The flower consists of three petals that curl around to form a sac with a long spur. Due to the length of the spur, only hummingbirds have the ability to reach the nectar. As these tiny birds enter the flower, pollen becomes attached to their long bills and they carry it off to the next flower, thus facilitating pollination. In early to mid-September, ruby-throated hummingbirds aggressively feast on colonies of touch-me-not in wetland areas as they fatten up for their annual fall migration.

The long spur keeps bees and other insects from stealing the nectar. Where hummingbirds are uncommon, the plants have become self-pollinating.

photo: John Toren

Touch-me-nots thrive in the cool, moist environment of the Eloise Butler Wildflower Garden in Minneapolis.

White Snakeroot

White snakeroot is a seemingly harmless plant that grows along woodland edges speckled with sunlight. Its milk-white, composite flowers, which look like starburst clusters, appear in midsummer, standing out in the shaded dimness of forest trails. The Iroquois traditionally used the plant in sweat baths to cool a patient and as a poultice to draw out the poison of snakebites. Unconscious patients were said to be revived from the smoke of the burning flower. White snakeroot's less healthful effects began to emerge in the early 19th century, when rural settlers began to suffer from symptoms that included weakness, tremors, and delirium. Some of them died. Even horses and cattle that had ingested the plant would stagger around in distress. The toxin from white snakeroot was even present in the milk produced by cows that had consumed the plant. Thousands of people died from what was called "the trembles" or "milk sickness" after drinking it.

The most famous person to have died from this disease was Abraham Lincoln's mother, Nancy Hanks, who succumbed in 1818, when Lincoln was nine years old.

An Illinois doctor named Anna Bixby suspected white snakeroot was the culprit causing milk sickness. She collected the plant and fed it to a calf, which then developed the symptoms of the disease. By 1834 she had eliminated the disease in her area, but because she was a woman her theory was not widely accepted. It wasn't until 1928, when a USDA researcher discovered the oily, alcohol toxin tremetol in the plant, that the mystery was solved.

Today humans rarely contract the disease, though livestock still succumb occasionally, and we can appreciate the beauty of the white snakeroot as it grows among the flora of the forest.

photo: David Brislance

The Haunting Loon Call

Hearing the lonesome wail of a loon quavering over a moonlit lake on a summer's night is an experience never to be forgotten. The chorus of several loons calling back and forth in a tone somewhere between madness and hilarity is even more memorable.

In flight, the common loon sometimes emits a plaintive lament as a contact call to other birds, and its alarm call—a quivering, throaty tremolo—is familiar to city-dwellers through television shows and films such as *On Golden Pond*.

Loons are fish eaters, and their bones are heavier than those of other bird species, making it easier for them to dive—but more difficult to lift off into flight. Because of their heavy body weight, they need at least a 90-foot runway in order to take off from the water. Loons sometimes mistake dark pavement for water—an often fatal error, because they cannot take flight from land.

In England loons are called the great northern diver, and they regularly dive to 100 feet, though they occasionally descend to 240 feet. Penguins are the only birds that can dive deeper.

Loons' feet are set far back on their bodies, making them very clumsy on land. Hence they always nest near shore. By one proposed derivation, the word "loon" comes from the old English word "lumme," which means lummox.

During migration, loons can fly 70 miles an hour and cover more than 600 miles in a day. Aside from nesting activities, they spend their time in the water. Young loons ride on the backs of their parents for two or three weeks for warmth and protection but can dive short depths soon after hatching.

The Ojibwe believe the loon was the first creature of creation. The Delaware believe it led survivors to land after the great deluge. The Inuit have more than 30 names for the loon. A loon is depicted on the Canadian dollar coin called the "loonie."

Yarrow

With its tiny, flat-topped white flowers and feathery leaves, yarrow could be considered the "plain Jane" of the floral world, and it's often overlooked or taken as just another weed. Yet it's one of the most widely used and potent medicinal plants in the world. It has been found in Neanderthal burial sites dating back 60,000 years.

The genus name for yarrow, *Achillea*, derives from the Greek hero Achilles who, according to legend, treated his wounded fellow soldiers with the plant during the Trojan War. It was still being used to stop bleeding from battlefield injuries during the American Civil War, when it was commonly known as soldier's woundwort.

Unlike most composite plants, the flowers of yarrow are tightly compacted, increasing the chance of pollination by visiting insects. It blooms for a long period, from May until the first frost, extending the time for pollination. In winter the seeds are dispersed by the wind.

The species name of yarrow is *millefolium* which means "a thousand leaves." In Colorado and New Mexico yarrow is called "plumajillo," which is Spanish for

"little feather," so named because of its delicate fern-like foliage. Yet yarrow is a tough, hardy plant that can survive in poor soils and during all kinds of weather conditions.

More than a hundred biological compounds have been derived from yarrow, including a number that have anti-inflammatory properties. It has traditionally been used by Native American tribes to treat colds, fevers, gastric problems, burns, anorexia, and infections.

It has a strong fragrant camphor odor that repels insects and helps protect nearby garden plants. It contains the chemical thujone, which is in absinthe. Addiction to this chemical might have contributed to the insanity and death of Vincent Van Gogh.

Yarrow was named "seven years love" because when a new husband ate the plant at the wedding ceremony, he would supposedly not leave his bride for at least seven years. It was also considered an aphrodisiac and called "old man's pepper."

In Europe yarrow was sometimes hung over cribs to keep witches from stealing babies.

photo: Carol Knabe

The Ruby-throated Hummingbird

Ruby-throated hummingbirds look like energetic fairies as they whiz and dart from flower to flower. This mighty-mite of the bird world needs to keep such a frenetic schedule to satisfy the fiery little furnace burning within. A hummingbird's heart beats more than 600 times per minute, giving it the highest metabolic rate of any vertebrate in the world—100 times faster than that of an elephant. At rest it takes 250 breaths per minute, and it beats its wings up to 80 times a second as it hovers in space near a tempting flower. It can dive at 60 miles per hour and often does so during mating displays. It is the only bird whose wings are attached directly to the shoulder joint; this explains why the hummingbird alone of all bird species can fly in any direction, including sideways and backward.

Ruby-throated hummingbirds weigh a tenth of an ounce—about the weight of a penny—and are three inches long, which is shorter than the toe of a bald eagle.

Hummingbirds prefer red and orange flowers, but they do not compete with bees for nectar because bees are color blind to those colors. Minnesota has only one species of hummingbird, a fact researchers attribute not to the cold climate but to the lack of red flowers. In the spring, before many flowers are in bloom, hummingbirds derive nourishment from the flowing sap from trees and the insects that feed on the sap. Hummingbirds have a fringed, forked tongue designed to hold liquid. They can flick their tongue three times a second while licking sap and nectar.

Only the male ruby-throated hummingbird has a ruby throat. Its iridescence flashes in the sunlight but might appear black at other times. After mating the female does all of the "housework," building the nest, incubating the eggs (which are the size of M&Ms) and raising the babies. It's like having a lazy husband around the house.

The hummingbird nest is about the size of a half walnut shell, and though seldom seen, it's a marvelous construction of plant down, attached to a limb with spider silk, and camouflaged with lichen.

Hummingbirds survive colder nights by going into a hibernation-like state called torpor, during which their body temperature can drop from 105 degrees down to 50 degrees, which lowers the energy level needed to keep alive. Before crossing the Gulf of Mexico during migration they double their fat mass, which makes it possible for them to survive the 500-mile non-stop flight.

photo: David Brislance

The Mayans believed when the Great God of creation made the birds, he had parts left over and decided to use them to make the smallest bird in the world, the hummingbird. Hopi and Zuni tribes believe the hummingbird has a direct connection to the rain god. Their water jars are artistically decorated with humming birds.

Find a patch of jewelweed in early September and you might be treated to a long parade of hummers bouncing in and out of the nectar-filled spurs of the flowers.

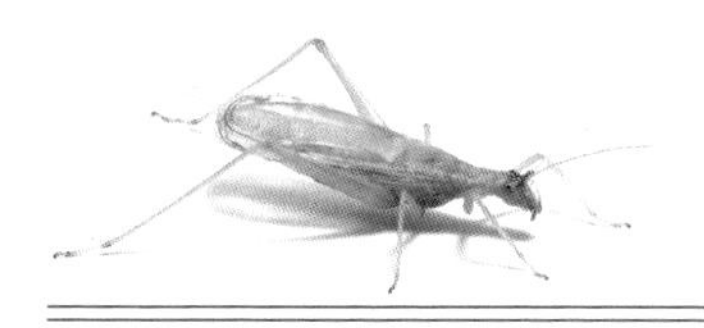

Snowy Tree Cricket

When the broiling days of July lead to languid, steamy nights, the snowy tree cricket, hidden in the bushes and shrubs of many backyards, begins its annual serenade.

Snowy tree crickets are pale green in color, with transparent wings and a blotch of red on their head below the antennas. Like others members of the family, male snowy tree crickets rub their wings together to produce sound. Females have hearing organs that can detect a much wider range of frequencies than most other insects, which increases their chances of locating a mate. Once they've mated, the female lays her eggs in holes she's drilled in the branches of a bush or tree. The young hatch in spring and begin feeding immediately, mostly on aphids. They can undergo ten molts or more before reaching adulthood.

Cheyenne buffalo hunters believed that chirping crickets would lead them to the herds, and many western tribes traditionally considered crickets chirping in the home as a sign of bad luck.

The snowy tree cricket is also known as the "temperature" cricket. By counting the number of chirps the cricket makes in 13 seconds and adding 40, you can determine the outdoor temperature to within a degree or two.

The open-air concert of snowy tree crickets is a nightly summertime musicale given free of charge. Enjoy!

Dandelions

Pictured on the next page is a packet of dandelion seeds distributed by the Northrup King Company in 1903. The seeds then sold for 10 cents. The packet, from my collection, informs the prospective buyer that the plant has "improved thick leaves" and is "excellent for home and market gardens." It's hard to believe that since those days, millions of dollars have been spent in vain trying to eradicate the dandelion.

Dandelions are high in nectar and have strong ultraviolet reflectance, which explains why they attract nearly 100 species of insects, including honey bees. They have one of the longest flowering seasons of any plant. In Minnesota, it's not unusual to see dandelions first flowering in March and thriving into November.

Dandelion root systems can penetrate the ground to a depth of a foot and can reproduce from roots and root fragments. They can also reproduce asexually and through self-pollination.

They are a composite flower with about 200 seeds on every seed head. When the flowers turn to seed, the stem grows higher in order to catch the wind, which disperses the seeds farther. When they're mowed, some plants become shorter in order to survive. That's evolution in action.

Dandelions produce more vitamin A than carrots and more potassium than bananas. Outdoorsman Euell Gibbons, a proponent of natural diets in the 60s and 70s, said the roasted root was the best coffee substitute in America. Recent studies have shown that an extract from the plant increases red blood cells, and in high doses, normalizes white blood cells. The root has been shown to stimulate weight loss and has antioxidant and diuretic features.

An old myth said that if children touched or ate the plant they would wet the bed... so it was not a potherb but a "potty" herb. In England children believed the seeds were fairies; if they caught one and made a wish it would come true—but they had to let the fairy go. If a maiden blew on the seed head, the number of seeds that remained would foretell how many children she would have.

It's ironic that the dandelion, whose flower is a symbol of love, has become an object of hate to so many.

photo: John Toren

Scott Mehus, education director at the National Eagle Center, and Angel, the center's bald eagle ambassador.

The Bald Eagle

Before Europeans arrived in North America as many as half a million bald eagles may have called the continent home. By the 1950's there were only 415 nesting pairs left in the lower 48 states. Eagles were shot because it was believed they preyed on young livestock and even carried away children. Alaska still had a $2 bounty on eagles as late as the 1950s. As if this wasn't enough, the indiscriminate spraying of DDT and misuse of other chemicals further decimated the eagle population.

In my hometown of Wabasha, Minnesota, we noticed the first eagles returning in the 1970s, a few years after DDT was banned. Little did we know then that Wabasha would become the site of the National Eagle Center, housing an eagle memorabilia collection of 20,000 pieces. In recent years this bluff-lined stretch of the Mississippi has become the premier viewing site for overwintering and migrating eagles in the lower 48 states.

More eagle nests—the largest nest any bird makes—are now being seen throughout Minnesota. Eagles mate for life and return to the same nest every year, adding new material to the old home, which can eventually approach a ton in weight. Female bald eagles are 25 percent larger than males and do most of the incubating, though both parents feed the young. An eaglet can eat as much as six ounces of food a day. Once fully grown, an eagle's powerful yellow feet have a gripping power 10 times greater than that of a human. It can typically carry up to four pounds back to the nest, with one record of an eagle carrying a 15-pound fawn.

Eagles will sometimes catch a victim in the water and be unable to lift it. I once saw an eagle on Lake Superior swimming to shore and presumed it must be injured. When it reached land I realized it was dragging a herring gull. The eagle proceeded to pull the unfortunate critter up on the rocks, pluck it, and eat it.

The Choctaw consider the eagle a symbol of peace, and many tribes use eagle feathers for ceremonial purposes. On the less positive side, eagles also rob prey from other birds, especially ospreys. They are considered klepto-parasites, and Ben Franklin considered the bird to be a "robber of bad moral character" and a "rank coward." The bald eagle was nevertheless voted by Congress to be our national bird over the wild turkey.

photo: David Brislance

photo: John Toren

Wetland Summer Walk

A trip to a wetland boardwalk on a lush summer day, when cumulus clouds and gentle breezes temper the afternoon heat, can be a soothing experience. Seated on a bench at the boardwalk, we can sit quietly and let nature come to us. The plant life is now in full bloom. The bur-reed, with its spiked green flowers resembling miniature hand grenades, is sharp to the touch.

The plant with tiny white flowers shooting out in whorls like fireworks on a Fourth of July night is called water plantain. It may seem that a novice bluegrass musician has begun practicing the banjo nearby, but that "plunking" sound in the water is actually the mating call of the green frog. The tiny gray tree frog, with its suction-cup feet, clings to the broad, pointed leaves of the arrowhead near shore, though its chameleon-like ability to change colors usually keeps it well-hidden from view. The genus of arrowhead is *Sagittaria*, after Sagittarius the archer of the summer sky. Also called *wapato*, arrowhead tubers are edible and were traditionally dried for winter food by some Native American tribes.

Barn swallows and tree swallows swoop and dive above the water, consuming large quantities of pesky insects. A hooded merganser and a mallard mom furtively glide in and out of the cattails and sedges, each followed by a brood of little paddlers that no one would deny are cute. When looking for food, the mergansers are scuba divers, while the mallards are snorkelers, with their heads under water and their bottoms in the air.

A male hooded merganser

The ruby, cherry-faced, and yellow-legged meadowhawk dragonflies are a few of the many dragonfly species that occupy these marshy domains. Males flaunt their resplendent red colors, while females and juveniles dress in more muted amber tones. As they flit on the wing in the warming afternoon sun, there is no way for us to predict which direction they'll take next—up, down, forward, or backward.

A muskrat sits on a feeding pad quietly nibbling its favorite cattail snack. If alarmed, it will dive under the water and reappear as far away as possible from the perceived threat. Cattails are not only the favorite food of muskrats, but also the main construction material for their floating houses. By keeping cattails under control, muskrats play a key role in preserving open wetlands in which waterfowl and a variety of other species can thrive—and we can enjoy watching them go about their business.

Tyrant Flycatchers

Among the several subgroups of flycatchers that spend their summers in our region, the tyrant flycatchers deserve a special note. This family includes the eastern kingbird, the great crested flycatcher, the wood pewee, and the phoebe. They've been given the name "tyrant" because of their aggressive behavior toward other birds, and "flycatcher" because they feed by flying from a perch, nabbing their meal in midair, and often returning to the same branch. This flitting back and forth while feeding is an easy way to identify the flycatcher family.

The question remains: which flycatcher is it?

The eastern kingbird is the most pugnacious of the four. It will chase crows and hawks high into the sky to defend its territory. I have been hit in the head more than once by an irate kingbird checking its nest. The kingbird has a dark gray back and white breast, but the white band at the end of its tail is a dead giveaway. It prefers open fields and orchards and can often be seen perched on a fence post or railing. Kingbirds often lays their eggs, creamy white and splotched with brown, in apple trees.

The great crested flycatcher, on the other hand, is more often heard than seen. A bird of the treetops, it can be identified by its long rusty tail and its loud whistling *whelp*, which might almost be compared to an avian war whoop. It's the only flycatcher that nests in tree cavities.

The nest often includes a cast-off snake skin. The skin may be used to ward off intruders, or maybe the birds just like the look of it. No one really knows. Three to six creamy-white eggs, thickly streaked with reddish-brown, fill the nest. If you take a walk along a woodland trail in midsummer, you might hear the vociferous *wheep* of the great crested high above you.

The phoebe, a smaller bird, returns as early as late March with the very first spring arrivals, a group that also includes robins, bluebirds, and killdeers. Since phoebes are insect-eaters, this early return can be dangerous if unusually cold weather persists. (To learn more about the phoebe, see page 30.)

The wood pewee is one of the last birds to return to our region—usually in the second week of May. It resembles the phoebe superficially in appearance, but you aren't nearly as likely to see one. It prefers deep woods and is more easily recognized by its song, a plaintive up-and-down *pee-a-wee* followed, after a long pause, by a discouraged two-note *pee-oo*. The wood pewee's nest is more like a hummingbird's than another flycatcher's. Usually built far out on a limb with bits of lichen, bark, and spiderweb, it contains three or four creamy-white eggs with brown spots on the large end.

In the silent stillness of a hot summer afternoon, the sweet sound of the pewee can travel a long way through the shadows of the forest.

Clockwise from upper left: great crested flycatcher, eastern wood pewee, phoebe, and kingbird

Ed Dick heading to town with two hawks and fox on board

Ed Dick's Story

I often think about how much society has changed since the 1950s, when I was growing up on a family farm in Wabasha, Minnesota. In those days we were close to the natural world every day. (Perhaps farm kids still are, though there are a lot fewer of them around.) This thought came to mind again when I recently reconnected with my cousin, Ed Dick, who also grew up on the edge of Wabasha, and we started to share stories about the wild animals we had as pets.

I recalled the evening my brother Ed found a baby skunk while bringing home the milk cows from the pasture. He took it to the veterinarian, who de-scented it—that would be illegal today—and it became a pet. Our little skunk was always playful and would follow us around the farm. He ate his meals with our numerous cats, and they all tolerated each other. We also had pet raccoons which were wonderful playmates. They were constantly investigating things, and their curious antics always made for interesting times. Eventually we returned them to the wild.

Ed Dick grew up in a south Wabasha neighborhood called "cow town" because his father pastured a couple of milk cows there in a grassy area. Today the Dick residence would be called a hobby farm. As a youth, Ed raised two red-tailed hawk babies that had been orphaned in the woods near our farm. He fed them red meat scraps from the local butcher shop. He would feed them in the morning and they would fly around during the day. They would return in the evening and wait to be fed alongside the family's rat terrier, Patsy, before they went into the chicken coop to roost with the chickens for the night. Ed would sometimes tie the two hawks to the handle bars of his bike and ride around town with them, always getting curious stares. Sometimes the hawks would sit on the railing leading to the outhouse of their neighbor, an elderly woman who would summon Ed to remove them so she could go the bathroom. He had the hawks for five years, but one day one of them failed to return, and the other one disappeared a few days later.

One day Ed's brother Ervin found a red fox kit near the railroad tracks and brought it home. Ed named the fox Bosco—also the name of a chocolate syrup popular at the time. Bosco slept in a box at night but would bark until Ed came to put him to bed. The fox eventually developed a taste for the family ducks, and Ed had to start tying him up.

St. Felix, the Catholic school in Wabasha, held a field day every spring where students ran races along with other activities. Ed brought Bosco along to show the children, and he was the hit of the day. Bosco seemed to enjoy all of the attention.

In the 1950s St. Felix also sponsored an annual homecoming parade and football game. One year Ed and his brother Donnie spruced up their model 'A' Ford and had the hawks perched behind them with Bosco riding in the backseat. The football game that year was played against the St. Agnes Aggies. They painted a sign on the side of the model 'A' that said "Fox the Aggies."

How fortunate we were to live in a time when wild animals provided so much fun and entertainment. Those experiences taught us to respect and love the natural world.

photo: John Toren

The Mighty Oaks

The gnarled and misshapen branches of an aging oak look like the crumpled hands of a centenarian. And like centenarians, the ancient oak has weathered many storms and hardships, and yet retains a dignity that belies its years.

White oaks often live for 200 to 300 years, though a few have reached 600 years of age. Without such longevity the species probably would have vanished long ago, because it takes 50 years for a white oak to produce acorns. (Acorns of white oaks mature in one year; those of the red oak take two.)

Many animals compete for acorns including deer, squirrels, bears, and turkeys. Turkeys can hold up to 50 acorns in their craw. Squirrels eat white oak acorns as they ripen but often bury the more astringent red oak acorns. Red oak acorns need to be buried in order to successfully germinate, and squirrels play a key role in the propagation of this tree. White oak leaves are lobed, distinguishing them from the red oak leaves, which are pointed.

Many Native American tribes consider the white oak a source of strength and protection. Older trees are considered sacred and spiritual ceremonies have often been performed in their shadow. Many tribes have also considered the oak a medicinal tree. A tea from the bark was used to treat dysentery, bleeding, and as a gargle for sore throats. The Penobscot ate the acorns to induce thirst and foster a healthier lifestyle. Numerous tribes ground acorn kernels into a meal to produce hard bread, especially in California. White oak acorns were once used to treat cancer, and experiments have shown they have anti-viral and anti-tumor tannins.

Wasps and other insects lay eggs on oak leaves, which produces galls. An ink can be produced by mixing iron sulfate with the tannin in these galls, and such an ink was widely used from the fifth to the nineteenth centuries. Both Leonardo da Vinci and Johann Sebastian Bach used it, and oak ink was also used in the drafting of the Declaration of Independence and the U.S. Constitution.

photos: John Toren

Many recipes can be found online for making bread from acorns. Leaching the nuts properly is essential.

photo: David Brislance

A gray fox

The Fox

Red-coated hunters yelling Tally Ho! have given the fox a reputation for slyness and elusiveness, and anyone who grew up on a farm is likely to agree: the raucous cackling in the henhouse means that a fox is searching for a chicken dinner. The fox's intelligence and cunning figure prominently in folklore, though these qualities are usually given a slightly underhanded sheen. In Aesop's fable "The Fox and the Stork," for example, the fox invites the stork for a meal of soup that the stork cannot consume because its bill can't hold it. The stork returns the favor by offering the fox a meal in a long jar that the fox can't reach with his thick snout. The moral of the story is, "One bad turn deserves another."

A member of the dog family, the fox is gifted with excellent hearing and smell as well as keen eyesight. The average fox weighs between 6 and 15 pounds, but because of its long bushy tail and rangy body, it appears to be much larger. Fox tracks are usually seen running in a straight line, with footprints about the size of a house cat's but with distinguishable claw marks. When foxes are hunting, however, their tracks look more like a drunken soldier's.

The fur of the red and gray foxes are strikingly different. Trappers prize the soft, silky pelt of the red fox, while the coarse, bristly hair of the gray fox has little commercial value. Red foxes have "black socks" on their feet, while gray foxes sport a stripe of black hair that runs down the middle of their tail.

The red fox hunts both day and night. Its gray cousin hunts mainly at night. Rabbits are a staple in the diet of the gray fox, whereas mice are the number one food item for the red fox. Foxes will kill more than they can eat and cache what they cannot consume for a later meal.

A red fox

photo: John Pennoyer

The gray fox prefers wooded areas, has hooked claws, and is the only member of the dog family that can climb trees. It returns to its den to rest. The den may be underground or as high as 30 feet in a tree. The red fox prefers open country, where it will sleep curled up in the most severe weather, sometimes completely buried in snow.

Sadly, foxes are often afflicted with mange, an itching skin disease caused by a parasitic tick. Red foxes have more readily adapted to human intrusion and therefore have become more dominant than the gray fox in the eastern United States. Coyotes will kill foxes, so the foxes have learned to live in the niches between coyote territories.

It's a wonderful experience to watch these graceful creatures romp and play in the summertime grasses or winter snow.

The Meadow

An open grassland can look dull at first glance, but in June it's filled with the sights and sounds of nature. The meadow has become a lush green carpet sprinkled with a variety of colorful blooms. Daisy fleabane shows white, pink, or purplish rays surrounding a citron center. Legend has it that if the fleabane has white rays, a boy is on the way, and if pink, a girl can be expected. The nearby hairy-stemmed black-eyed susans, with their cheerful yellow rays, are intermingled with the lavender blooms of alfalfa, and bumblebees' legs become heavy with pollen as they pry into the flowers of the red clover.

Attempts to introduce clover to New Zealand failed until bumblebees were also introduced, after which time the clover began to thrive. The bumblebees were the only insects strong enough to penetrate the deep blossom of the clover and ensure pollination.

Smooth bromegrass, with its large silky panicles, sways with every puff of wind. The densely packed spkeheads of Timothy grass look like Fourth of July sparklers. This species was unintentionally introduced to North America by early settlers; a

photo: Carol Knabe

farmer named Timothy Hanson began to promote it as a hay source around 1720, and the grass has been known by his name since then. Timothy has become naturalized throughout much of North America.

Orange sulfur butterflies dangle from alfalfa blossoms as they nectar on their favorite flower. The orange-brown common ringlet, with its notable black eye spot surrounded by yellow, bounces with an uneven flutter as it feeds on fleabanes and hawkweeds.

June is the time of year to hear the most melodious songster of the eastern grasslands: the bobolink. Thoreau wrote, "Away he launches and the meadow is all bespattered with melody." The male bobolink looks like an upside down tuxedo—solid black below and largely white above. Another resident of the open fields is the eastern meadowlark, easily identified by his yellow breast, black bib, and chopped-off tail edged in white. Meadowlarks joyously call from a power line or fence post, offering up a liquid "spring of the year."

The songs of these birds are the real "oldies but goodies."

Passenger Pigeons

In the 1800s the passenger pigeon represented 25 to 40 percent of the land birds in America. Lined up beak to tail, these birds would have stretched around the equator 22.6 times. Docile, flavorful, and unprotected by law, the species became a favorite target of market hunters during the mid- to late 19th century. In fact, they were so heavily hunted that the entire species, which had once numbered between three and five billion individuals, was wiped out. The last known passenger pigeon, "Martha," died on September 1, 1914, at the Cincinnati Zoo.

In 1806 Alexander Wilson, the father of American ornithology, witnessed a passenger pigeon flock in Kentucky that he estimated at more than two billion birds. He judged that it had been roughly a mile wide and 240 miles long. Audubon saw a flock of similar size and compared it to an eclipse of the sun.

Passenger pigeons nested in vast colonies. One of the largest ever recorded was near the town of Petoskey, Michigan. It was estimated to be 28 to 40 miles in length, three to ten miles wide, and covering at least 100,000 acres. In nesting areas, the calls of the pigeons were like rolling thunder. Large branches were constantly crashing to the ground from the sheer weight of the birds. Pigeon droppings lay two to three feet thick on the ground.

Male pigeons would gather sticks and pass them to the female during nest construction. All the females in a colony laid a single egg on the same day. Both sexes incubated the eggs, with all the females incubating at one time and then trading off with the males.

The mass slaughtering of the passenger pigeon took place from 1840 to1880. The young pigeons called squabs were fattened with pigeon's milk, which was a regurgitated mix, rich in protein, fats, and salt, from the crops of the adults. The squabs were like little butterballs and were considered a special delicacy. Pigeon hunters would take a captured pigeon, cover its eyes or sew them shut, and then tie it to a stump or perch to act as a decoy for other individuals flying overhead. The term "stool pigeon" derived from this cruel method of deception.

Usually the great flocks would return to Minnesota in April during Pigeon Days, causing excitement among not only hunters but the population at large. In the 1870s one of the largest nesting sites in the state was just south of Wabasha, extending along seven miles of the bluffs near the Mississippi River and into the hills for half a mile. It was said that nearly every oak tree contained nests covering 2,000 acres, with an estimated population of several million birds. A quarter century later, the bird was gone forever.

The Cowbird

photo: Carol Knabe

The brown-headed cowbird is the only bird in North America that relies on other birds to raise its young. The female lays her eggs in the nests of other birds and lets the surrogate parents do the rest.

Other bird species may not like this parasitic approach, but it's well-suited to the wandering, hobo lifestyle of cowbirds, which evolved as they followed the buffalo herds that roamed the prairies. There are records of more than 220 different species playing host to cowbird eggs, and 140 species are on record as having successfully raised cowbird young.

A female cowbird can lay as many as 40 eggs in a season, but because a number of species have recognized the cowbirds threat, only 3 percent of the eggs hatch and reach adulthood. Yellow warblers bury the cowbird egg with new nesting material; nests have been found with six layers of material covering eggs—a bird's version of a high-rise apartment. Gray catbirds reject as many as 95 percent of cowbird eggs and will sometimes abandon their own eggs in order not to hatch a little parasite. Robins have learned to identify the parasitic egg and will cover it up or toss it from their nests.

Cowbirds have recently been filmed killing the young of the host bird, leaving only their own juveniles alive. A study by Florida's Museum of Natural History found that if a foster parent threw out a parasitic egg, the female cowbird would return and more than half the time would destroy the entire nest.

While cowbird parasitism has been blamed for the decline of a number of other bird species, it's more likely that habitat destruction by humans has had a greater negative impact on bird numbers. By fragmenting and eliminating eastern forests, humans have made it possible for these birds of the plains to expand their range into new territories. In spite of their parasitic nature, cowbirds are a great benefit to farmers, eating crop pests such as grasshoppers, aphids, and beetles. The role of the cowbird in the environment is complex and deserves more careful study.

Clouds

As youngsters on our bluff-top farm, my brother and I liked to lie on the grass and watch the cotton candy puffs of cumulus clouds drift across the sky, imagining them to be different creatures as their ever-changing forms passed overhead. These clouds were a sure sign of storm-free days and gentle winds. Some mornings we would awaken to a bleeding sun forcing its rays through a thick bank of soupy stratus clouds. It meant that stormy weather was on the way. On hot, muggy afternoons cumulonimbus thunderheads would ascend ominously in the southwestern sky, threatening to batter us with damaging winds and bolts of lightning. Most often, however, they brought only a welcome downpour cooling the thirsty, feverish earth.

After those thunderstorms passed and the sunlight started to peek through the clouds again, a rainbow would appear, shining its bright array of colors into the woods, revealing the exact spot where a pot of gold lay hidden.

Before radar and Doppler weather forecasting, farmers used their knowledge of cloud formations to predict the weather. Windswept high-ceiling cirrus clouds resembling mares' tails invariably foretold of rain to come, a prediction confirmed by the weathervane on our barn, which would suddenly swing to the east.

My father would explain to us that if the sun set in a "nest," rain could be expected the next day. The nest he referred to was a thick mass of clouds that would appear above the western horizon to meet the sun as it disappeared into the purple cloud bank.

In autumn, colder, violet cumulus clouds lined with silver would replace the creamy summer puffs as the wind shifted and began to arrive from a more northerly directions. These clouds were sent by Jack Frost to remind us that snowflakes were soon to appear. Even the drab days of December could be brightened up by the sundogs that sometimes appeared on either side of the winter sun. These halos form when sunlight interacts with ice crystals in the atmosphere. They can appear as a welcome display of beauty on a frigid winter day.

Watching the daily cloud formations can also remind us of our own lives—always moving, changing, and unpredictable.

photo: John Toren

Midsummer Twilight Walk

A walk at twilight down a country lane in mid-summer can be a soothing experience. This is the time when the damp coolness of the evening begins to temper the sticky heat of the day. The animals on the day shift are checking out and those on the night shift are signing in. A robin chortles his final notes of the day while his partner spreads her protective wings across her brood of naked babies now snuggled comfortably in their nest. Vesper sparrows—named for their even-tide song—flush from the grass, flashing their white outer tail feathers.

A doe steps onto the path and stomps her hoof as a warning sign to the two spotted fawns that follow behind. They silently leap into the underbrush and are gone, like apparitions. A chubby raccoon hustles by, probably looking for his buddies. They will all soon be fighting and snarling over the remains of some unsuspecting human's garbage bin.

As darkness descends, little brown bats appear, swooping and dive bombing around us as they hone in on the tiny, but nasty, mosquitoes. A little brown bat mother can eat up to 4,500 mosquitoes and other insects in a night.

A nearby pond resounds with the trilling of American toads. This deafening hullabaloo will end within a few days, as the toad's mating season winds down. In the tall grasses edging the water the *zt...zt...zt...zeeee* call of the common meadow katydid is our musical companion. This pretty, lime-green insect, with long antennae and yellow rear abdomen, sings both night and day.

How did the katydid get its name? As the story goes, a young woman named Katy fell in love with a handsome young man, but he married someone else. Not long after the wedding, the happy couple was found poisoned in their bed. Who did it? On warm summer nights, these insects tell us what they saw: "Katydid, Katydid."

On the eastern horizon a faint red light fades in and out just after sundown. It might be mistaken for the warning light on a distant airplane tower, but it's actually the red planet Mars, which some optimists believe will be the new home for the human species someday. The football-shaped gibbous moon appears to have been kicked high into the heavens by a Titan deity, sprinkling yellow stardust in its path across the Milky Way.

Taking in this peaceful scene is a nice way to end the day.

Fall

The Wonders of Autumn

For nature observers, autumn arrives several weeks before its "official" debut at the fall equinox on September 21st. The profusion of summer greens has slowly been replaced by splashes of rainbow hues, as if Mother Nature were growing tired of her verdant artwork and had decided to create a different backdrop for her upcoming performance.

The leaves of the walnut tree phase to a flaxen yellow, and squirrels vie avidly for the tasty nuts that cover the ground. The diminutive asters, our star flowers, display their tints of purple-blues and creamy whites until the killing frosts finally topple them. And the flaming vermillion foliage of the Virginia creeper acts as a fall foliar flag calling migrating birds to feed on its nutrient-rich blue-black berries. This vine has powerful tendrils, five of which can hold up to ten pounds and can stay attached to a surface long after the supporting plant has perished. A Virginia creeper which has enveloped a dead elm or oak tree looks like a thin scarlet bonfire.

This is the time of the year when the whistling calls of long skeins of tundra swans can be heard as they fly southward, blown from the northland by arctic winds. Canada geese are also on the wing, honking lustily in their familiar V-formation, apparently the only letter of the alphabet they know. Kettles of turkey vultures with their ebony bodies and silver under-wings circle overhead searching for the remains of any critter who happened to cross the highway at just the wrong time. These natural garbage disposals render an important service as they wend their way to hotter climes.

Though the dwindling days of sunlight are difficult to measure with the eye, autumn's advance can be marked by the last time we hear or see a certain species. When did we last see a monarch butterfly as it fluttered its way to its winter home in the Sierra Madres in Mexico? What day did the hardy little Carolina ground cricket give its final chirp? And when was the last time we heard the cheery call of the eastern bluebird?

As the seasons change we remember the old and await the new.

photo: John Toren

Early fall colors at Clifton French Regional Park in Plymouth, Minnesota

The Gingko Tree

I am fortunate to have a descendant of a 50-foot fossil in my backyard. Well, fossils don't really have descendants. And no, it isn't a dinosaur. It's a gingko tree, a species that was thriving 225 million years ago during the Paleozoic era, when dinosaurs ruled. The gingko is native to China but is now extinct in the wild. It's considered to be the oldest cultivated tree on earth and unrelated to any other species.

Only female gingkos bear fruit. We seldom see the flowers, which bloom only at night and drop immediately. The fruit of the gingko resembles an apricot in size and color and has an odor of doggie doo, which has earned it the folk name "stink bomb" tree. In spite of its rancid odor, the omnivorous, ubiquitous neighborhood raccoons have left evidence in my backyard that they consume the fruit.

Gingko translates as "white nut" in Chinese, and the edible nut has been used for centuries in Asian cuisine. An old folk belief was that if a female gingko was planted near a pond, the reflection from the water would make it fertile.

Many cities are now planting gingkoes because they are tolerant of pollution, cold hardy, disease resistant, and grow well in poor, compact soils. When young, the branching structure is usually awkward, but in time the gingko assumes a pleasantly rounded shape. In autumn the elegant fan-shaped leaves turn a sun-blazed yellow and drop within 24 hours—nature's version of time-lapse photography.

Mushrooms

The late-summer showers of August and September foster a fungal eruption in the moistness of shady woodlands, presenting those who wander the forest paths with mushrooms in a dazzling array of colors, shapes, and sizes.

The ivory white surface of the giant puffball stands in stark contrast to the dark forest floor like the bottom third of a melting snowman. It can weigh more than 50 pounds and grow to five feet in length. The doughy inside is protected by a leathery skin that eventually ages and cracks, dispersing trillions of olive-gray spores upon contact with wind or rain.

At the other end of the spectrum lies the gem-studded puffball, which is about the size of a ping-pong ball. It's surface is studded with tiny white spikes that give it a jewel-like appearance. As it ages a vent hole appears at the top, and when squeezed, the spores rise up like smoke from a chimney.

The pumpkin-orange Jack O' Lantern grows in large clumps, and both the fruiting body and the mycelium (roots) give off a green glow due to a chemical reaction within the mushroom involving a protein, oxygen, and calcium. People are often startled by this phenomenon, called bioluminescence,

Jack O' Lantern mushrooms

which can be observed on dark nights. Gardening friends once told me they had left their wheelbarrow filled with debris as night approached, and looking out in the yard later were amazed to see that the contents of the wheelbarrow were aglow with a shining green light. They were happy to learn it was not the work of a hobgoblin but that of a Jack O' Lantern mushroom.

The Tippler's Bane mushroom, a member of the inky cap family, is well-named. It's edible, but those who drink alcohol two to four days after eating it may experience numbness in the face and hands, rapid heartbeat, nausea, vomiting, and other symptoms. Explanation? This mushroom has a unique amino acid called coprine that stops the metabolism of ethyl alcohol. While it doesn't kill the unlucky diner, it will give them a hangover for the ages.

Minnesota has more than 5,000 mushroom species. Searching for and finding new kinds is like uncovering hidden jewels tucked away in the shadowy forest, but positive identification is often difficult. The best and safest way to approach this vast field of organisms is in the company of an expert.

photo: Nino Barbieri

Shaggy Mane mushrooms

Wild Turkeys

photo: John Pennoyer

One of the more stellar attractions we might be lucky enough to come upon in springtime is the rainbow arc of the tom turkey's tail feathers. Such a sight would not have been likely in Minnesota, however, until 1971, when the DNR acquired a flock of 13 turkeys from Missouri in exchange for some ruffed grouse. The 13 were released in Houston County and started the successful return of wild turkeys to the state. Earlier attempts to reintroduce the bird had been failures. In 1923, the state game farm at Mound raised birds from Maryland, Pennsylvania, and Texas, and released 250 of them in 13 counties, but none survived, largely due to raccoon and great horned owl predation. Efforts to introduce the *Merriam* subspecies from western states had also proved futile, but the larger, hardier eastern subspecies from Missouri that was released in 1971 was soon thriving.

In 1932, Thomas Roberts wrote in his pioneering study *Birds of Minnesota* that there were no credible reports of turkeys ever being in Minnesota. But more recent research suggests that turkeys were a native species. Roberts also stated that if turkeys ever existed in the state, it could only have been in southeastern region, but "these areas are now so thickly settled that it seems unlikely that turkeys could exist there under present conditions." Even experts can get it wrong sometimes.

By the 1930s, due to overhunting and loss of habitat, the wild turkey population in the U.S. had been reduced to only 30,000 birds. With hunting restrictions and habitat improvement, turkey numbers have rebounded to an estimated 7 million and are now hunted in every state except Alaska.

Turkeys are native to North America but were taken to Europe by the Spanish and reintroduced by the Pilgrims in the 1600s. English settlers thought the birds were a kind of guinea fowl from Turkey—hence the name.

Native American tribes domesticated the turkey some 2,000 years ago, more for the feathers than the meat. Some modern tribes consider the turkey an inferior

photo: John Toren

Turkeys have become a common sight in many suburban neighborhoods.

bird and a coward, and they refused to eat the meat for fear of becoming cowardly themselves. But some tribes considered them spiritually important and buried them with ceremony when they died. Turkey feathers were attached to prayer sticks given as an offering to the dead. The Pima consider the turkey to be a rain spirit that can predict the weather. Pioneers used the spurs from the legs of turkeys as arrowheads for shooting small game.

An adult bird has about 5,500 feathers in all. Tail feathers were used as fans, wings were used as brooms, and soft downy feathers were used as stuffing for mattresses and pillows. The Fremont Indians of central Utah made exquisite blankets out of turkey feathers.

Turkeys can run up to 25 m.p.h. and have a flight speed of 55 m.p.h. They are omnivorous but 80 percent of their diet consists of grass. Dandelion flowers are a major food source in springtime.

I have found turkey nests along woodland trails and in tall grass fields. They can lay up to 20 eggs. The largest nest I have found contained 18 eggs. Young turkeys are called poults. Year-old males are called jakes and females jennies. Turkeys roost in trees at night, moving from site to site in an effort to avoid their nighttime nemesis the great horned owl.

photo: Jim Gilbert

Witch Hazel

Witch hazel is typically the last plant to bloom in Minnesota each year. Its yellow, four-petal flower clusters, which resemble tiny lightning bolts, begin blooming in September and continue through November. This extended bloom period helps ensure that the few insects still active, mostly small gnats and wasps, will locate the flowers and complete pollination.

The genus name *Hamamelis* means "together with fruit" and refers to the simultaneous occurrence of the flower and the mature fruit. When the witch hazel blooms, the seed pods from the previous year begin exploding like Orville Redenbacher's popcorn, shooting the two black seeds in each pod 10 to 20 feet. The term "witch" is from the Old English meaning "to bend." The forked branches of witch hazel are still used, as they have been for centuries, as divining rods to locate underground water sources. The rods supposedly bend whenever water is near.

In the 19th century a home remedy was popular consisting of volatile oil from the witch hazel mixed with alcohol and water. In 1846 Dr. Charles Hawes developed a system of steam distillation for the plant twigs and used it to brew "Hawes Extract," the first commercial use of witch hazel by this method. The Ponds Extract Company called a similar concoction "the people's remedy" for burns, scalds, bruises, sprains, lame back, frozen limbs, broken breast, rheumatism, bleeding piles, toothaches, sore throat, bleeding lungs, sting of insects, and hemorrhages.

The annual witch hazel bloom can be seen in the Dayton Wildflower Garden at the U of M Landscape Arboretum. The burst of yellow flowers adds color to the drab gray background of November.

photo: John Toren

Tens of thousands of hawks pass above Hawk Ridge in Duluth every autumn on their migration to warmer climates.

Bird Migration

How do birds find their way during migration? It's an age-old question, though scientists are coming closer to finding some definitive answers. Studies have shown that birds use landforms, stars, the sun, the earth's magnetic field, and odors to guide their path. Which method is put to use varies among species. Experiments with indigo buntings using a planetarium's star ceiling showed that the buntings found the proper magnetic direction based on the "sky map." When the planetariums stars were shifted, the birds shifted their orientation as well. Pigeons have shown they are able to detect the earth's magnetic field. We still do not fully understand this phenomenon or know if other species of birds have the ability to do this. Birds also watch the sunset to keep properly oriented at night.

Birds migrate to their northern breeding grounds to avoid competition in their wintering areas and to take advantage of the insect-rich food supply and longer daylight, which provides for more feeding time and faster development of the young. Some early fall migrants head south from the Arctic as soon as the first week in July. Included in this group are the greater and lesser yellowlegs and the sandpipers, which leave their young in their Canadian breeding grounds as soon as they're able to fend for themselves. This is not bad parenting; once the adults leave there will be more food for their offspring.

Every spring about 150 species of land and freshwater birds migrate from the southern U.S. and Central and South America to breed in the north. The migrating arctic tern can cover 11,000 miles from Antarctica to the Arctic. This tern species likely sees more daylight hours than any other species on earth. The most rapid shore-bird flight documented by banding was that of a lesser yellow-legs that flew from the Massachusetts coast to the island of Martinique, traveling 1,930 miles in just six days! That's 322 miles per day—not bad for a bird that weighs only 3.5 ounces. In contrast to these marathoners, the Clark's nutcracker, a western species, moves only a few miles up into the mountains every spring to nest. This is called altitudinal migration. Most long-distance migrants fly at night and stick close to the ground during daylight hours to rest and feed. Aerial foragers like swallows and swifts fly non-stop, feeding along the way.

Passerines (perching songbirds) lose one-fourth to one-half of their body weight as they migrate long distances over water. Needless to say, storms can have devastating effects on their progress. On the night of March 13, 1904, in southwestern Minnesota and northwestern Iowa, millions of Lapland longspurs were killed, flying against buildings, electric light poles, and wires, and forcibly driving themselves onto the frozen ground and ice during a very dark night with wet, heavy snow. On two small lakes in Worthington alone, an estimated 750,000 birds perished. One hundred of the bodies were dissected; all had died by violence, mainly crushed skulls, broken necks, and internal hemorrhages. Their stomachs were empty, which could have contributed to their demise. Amazingly, within a few years the population had rebounded.

Migration is the most exciting time for bird watchers to be afield. Many of our feathered friends are returning north after a long absence, though like old human friends, they're sometimes so quickly gone that we miss them entirely. Migrations, like life's passing days, literally fly by. Let's stop a while and enjoy them.

The distinctive call of the nighthawk as it migrates south across Minnesota in late summer is an annual treat.

Bird Nests

There are a number of ways to identify birds in the field, including color, size, song, and flight pattern. Another is to examine the bird's nest. Each bird builds a nest unique to its species. Learning to identify nests is a wonderful way to discover more about birds and the important role they play in the environment. It also helps us recognize which habitat a particular bird prefers, and sometimes why.

Birds typically use grasses, leaves, and twigs to build nests. Some nests are loosely constructed of large twigs and unlined. Others are constructed of pine needles and insulated with feathers, moss, or dead leaves. For example, eastern phoebes and chickadees line their nests with moss, while great blue herons and great egrets use sticks and twigs exclusively for nesting material. Barn and cliff swallows fashion their cave-like nests mostly of mud.

But sometimes birds find unique and amazing items to adorn their homes. It's not unusual for Baltimore orioles to use cord string to fasten their nests securely around branches in artisan fashion. (The poetic words "Rock-a-bye baby in the treetop, when the wind blows the cradle will rock" were written about the baby Baltimore oriole, not a human baby.) Robins, cardinals, house sparrows, and catbirds are some of the birds that will use pieces of plastic, landscape fabric, yarn or other human-made fabrics for nesting material.

At the U of M Landscape Arboretum, I once made an unusual discovery: a brown thrasher had decorated its nest with a dozen aluminum identification tags gathered from a pile of discarded research trees. Among other examples of strange material used by birds around the world, a Chihuahuan raven was found to have built a nest entirely of barbed wire (ouch!), and a colony of double-crested cormorants used pocketknives, pipes, hairpins, and ladies' combs taken from a sunken vessel. One western kingbird built a nest with a large collection of cigarette butts and filters—a bird with a smoking habit, no doubt. The author John Steinbeck found three shirts, one bath towel, one arrow, and his rake in an osprey nest built near his garden. Other items also found in osprey nests include a toy boat, old shoes, a rag doll, a straw hat, a broom and barrel staves. Clearly birds can be very creative when building a home. Let's help them out whenever we can.

By preserving and improving nesting habitat for birds, we're helping to protect our crops, trees, and native plants from insect pests. Birds also consume tons of weed seeds that, if left to germinate, would be chemically sprayed or removed in other costly ways. Whenever possible we should leave dead trees standing, as they provide nest sites, food, and protection for at least 69 species of birds and animals in Minnesota. By not mowing fields until August we allow young birds time to fledge. Early mowing of hayfields has had a devastating effect on many bird species including meadowlarks and bobolinks.

photo: John Toren

Building nest boxes for cavity nesting birds is another way to improve the chances of survival for a number of species. The bluebird recovery program has been a tremendous success. Bluebirds had become endangered before a nationwide effort was started to construct houses and place them in suitable habitat. As a result, the bluebird population rebounded and once again stands at healthy levels.

Bittersweet turns orange on the vine in autumn, adding a splash of color to the fading vegetation.

Mother Nature in Autumn

In mid-October, when woodland trails are sprinkled with the golden-yellow leaves of cottonwood, hickory, and maple trees, they can almost look like the Yellow Brick Road to Oz. Chipmunks chitter and scurry about, acting as if we're intruding on their efforts to store up provisions for the snowbound months ahead. In the underbrush, white-throated sparrows scratch about for weed seeds and other food bits in an effort to fatten up for their annual trek to milder climes.

In the wetlands, muskrats are busily constructing their floating houseboats of sedges and cattails. Without these industrious rodents the marshes would be filled with vegetation, eliminating watery stopover areas for ducks, geese, and other waterfowl.

November is the month of clouds drifting through a tangerine sky—the puffy, purple-plum kind with silver-tinseled lining. The skyline is filled with the murmuration flight of starlings—a complex weaving pattern of countless birds in a shimmering restlessness that resembles the canvas of a pointillist painter working in black and white. Lengthy strands of snow geese pass high overhead, emitting the shrill, nasal *houch, houch* chorus that often reaches our ears long before the birds appear.

Pumpkins now lay in the fields, looking like orange meteorites on an otherwise barren moonscape. Those that are not harvested become a favorite snack for white-tailed deer. On warmer afternoons, the melodic chirping of the fall field cricket sounds the concluding notes in the annual insect symphony. Though similar in appearance and sound to the spring field cricket, the fall field cricket is a separate species with a different life cycle.

The fading colors and sounds of fall represent Mother Nature's last sigh of activity before she settles in for her long winter's nap. One morning we wake up to a white panorama, and it almost seems that we've been transported to another universe.

So go the seasons of the year, and of our lives.

photo: John Toren

In late November, ice begins to form on the lakes at Lebonon Hills Regional Park.

Tamarack

There are 826 species listed on the National Registry of Big Trees. Minnesota holds the record for only one of them, the tamarack (*Larix laricina*). This monster is in Crow Wing County and is known as the Witch Tree. It's 71 feet tall and 133 inches in diameter.

Tamarack is the only deciduous conifer in the state. In other words, unlike most cone-bearing trees, the tamarack drops its needles every year. Every spring soft blue-green needles in bundles of 10 to 20 emerge from knobby protrusions on the tree's branches. In autumn these needles take on a golden hue that illuminates the sometimes monotonous bog landscape. The trees may look diseased at this stage, standing in front of a thick stand of dark green spruce, but when the tiny red-rose cones appear in spring it becomes clear that they're alive and well.

Tamaracks prefer the acidic bogs and swamps of northern Minnesota but will grow as far south as the Twin Cities. A large stand can be found across the parkway from Eloise Butler Wildflower Garden in Minneapolis. But tamaracks also grow farther north than any other tree species in North America, surviving to the edge of the tundra.

Tamarack is an Algonquian word that means "wood used for snowshoes." The wood is extremely rot resistant, and it was used for wigwam poles by several Native American tribes, while the roots played an important role in the manufacture of birch bark canoes. The resinous gum of the tamarack was chewed for indigestion, and a tea was used as a gargle for sore throats. A poultice made from the inner bark was used to treat frostbite.

Less glamorous and statuesque than white or red pines, the tamarack nevertheless exhibits distinctive marks of character that make it well-suited to Minnesota's often boggy terrain.

Bald-faced Hornets

As gusty autumn winds gradually strip the trees of their leaves, they expose the gray, papery, football-sized nests of bald-faced hornets that have been hidden in the foliage all summer. These tattered remnants cling to the trees or shrubs they were glued to in the spring, and foraging birds continue to ravage them looking for insects, but by this time of year the nests have been abandoned and all the hornets except the fertile queens are dead.

A tattered bald-faced hornet nest with a few of the insulating layers exposed

The bald-faced hornet has a black body with a cream-colored face and tip of the abdomen, giving it a menacing appearance that is well deserved. Whenever an intruder gets within three feet of their nests hornets become extremely aggressive, defending the hive. Unlike honey bees, hornets have smooth stingers that do not get embedded in their victims and therefore can sting multiple times.

After hibernating through the winter, the new queens emerge in the spring and begin to construct nests. The nest material is chewed wood fiber mixed with saliva.

The inner nest looks like honey bee combs with three or four tiers encased in the papery outer shell.

A colony of hornets can range from 100 to 400 members. To generate a population for the new year is a complex process. The queen lays hundreds of eggs, the first being sterile female workers who continue to build and maintain the nest. The queen fertilizes each egg as it is being laid by using sperm stored from the previous year. Occasionally she will not fertilize an egg, and it will become a male drone. Males have no stingers; their only purpose is to fertilize the queens. At midsummer the queen will lay eggs that produce new queens for the next year. At the end of the season the old queen dies naturally or is killed by the workers.

Bald-faced hornets are beneficial carnivores, ridding the environment of many aphids, flies, and other harmful insects. They also do some pollinating.

In some North American tribal myths, hornets are the creators of the Earth and a symbol of productivity and fertility. They can be pesky if aroused, but their nests should be destroyed only as a measure of last resort.

Black Walnut

As the daylight wanes and the green landscape of summer fades into the myriad hues of autumn, the flaxen yellow leaves of the black walnut trees add to the color bonanza. The leaves of the walnut fall early, along with the rough-skinned, husk-covered nuts.

Walnut trees produce the chemical juglone, a natural allelopathic herbicide that varies in effect from one species to the next. Black raspberries, forage grasses, and

photo: Ed Book

mints thrive on it, while tomatoes, potatoes, apples, pines, and walnut seedlings are unable to grow near it.

The wood from black walnut is considered to be the most valuable timber in the U.S. It's used to make fine furniture, veneer, and gun stocks. Dye from the husks was once widely used to color cloth, and older men would wear caps made of walnut husks in an effort to dye their gray hair. Carrying a walnut was thought to prevent rheumatism and cure insanity.

Squirrels have been responsible for spreading the walnut tree throughout the U.S. They bury the nuts but forget where they put many of them, thus advancing the walnut throughout the forest. Gray squirrels gnaw the nut from both sides, while red squirrels gnaw exclusively from one end or the other.

Archaeologists working in Pompeii, Italy, a city entirely destroyed by the Mt. Vesuvius eruption of 79 CE, unearthed stores of walnuts amid the buried homes and shops, thus proving that the "nut of the gods" has been enjoyed for centuries.

The Red-tailed Hawk

Red-tailed hawks often perch on light poles along highways, staring down from above like traffic monitors. In fact, they hardly notice the vehicles roaring past. What they're really interested in is pouncing on any careless mouse or vole that happens to scamper by. With eyesight two to three times more acute than humans, red-tails are able to identify the slightest movement in the roadside grasses beneath them.

Farmers used to call the red-tailed hawk a "chicken hawk," though the birds rarely prey on domestic fowl. Eighty-five to ninety percent of their diet consists of mice, voles, and other rodents. Keeping the population of these munching little critters in check is a real economic benefit to agriculture.

A red-tailed hawk can capture prey while perched or flying, but once its foot is locked onto the victim, the hawk can't release it until it lands on something solid, like the ground or a tree branch. Hawk is a Teutonic word meaning to "seize or take hold."

The red-tailed hawk is the most widely distributed and numerous of American raptors, with an estimated population of one million. They're monogamous, usually mate at three years and typically live to 12 years of age. They will often renovate old nests of crows or squirrels. When green sprigs of white cedar, hemlock, or pine are seen in a nest it signifies that it is active. Young hawks are called eyasses (pronounced EYE-ess-ez.) Immature hawks have gray tail feathers until the autumn of their second year, distinguishing them from the brick-red feathers of the adults. In old Western movies the screeching *kree-eee-ar* call of the red-tail usually meant something bad was about to happen.

photo: Sue Isaacson

Young red-tailed hawks are easily trained for falconry and are the species most in demand for this sport. About 5,000 red-tailed hawks are used for falconry in the United States.

In ancient Egypt hawks were considered royal birds. Ra, the Egyptian sun god and supreme deity, had the head of a hawk and the body of a human. The Polynesians believed the hawk was a prophetic bird with healing powers. In Greek mythology the hawk was the messenger for Apollo, the god of prophecy. In Iceland, if a person carried a hawk's tongue under their own it was said they could speak the language of the birds. Many Native American tribes treat the hawk as sacred and use its feathers for spiritual ceremonies and in headdresses.

Down through the ages hawks have been held in high esteem by many societies. We should treat them the same.

photo: John Pennoyer

The Ring-necked Pheasant

The male ring-necked pheasant is one of the most strikingly beautiful birds in Minnesota. Its head is a glossy greenish-purple with red wattles around the eye. The white ring around its neck is its most identifiable feature, while the multi-colored body and tail feathers come in hues of beige, red, and blue. The female has a long tail similar to the male, but her body is a muted blotched brown.

Originally from Asia, ring-necked pheasants were first successfully introduced in Oregon in 1881, and by the late 19th century they were flourishing in the eastern U.S. The first attempts to establish ring-necks in Minnesota, around 1905, were unsuccessful. It appeared that Minnesota winters were too harsh for pheasants to survive. More intensive breeding efforts were resumed in 1916 at the Big Island Game Farm on Lake Minnetonka. The farm was moved from Big Island to Mound in 1920 but was closed in 1928. In that 12-year span 25,000 pheasants were released throughout the state and 60,000 eggs were distributed to farmers for hatching.

Female pheasants brave the elements.

Although no ring-necks were released at the game farm, so many escaped that the colorful birds became a common site along roads and in yards and gardens around Minnetonka, and nearby farmers complained that escapees were causing significant crop damage.

The first designated pheasant hunting season took place in the fall of 1924. In 1926, despite an outcry from local residents, hunters received special permits to hunt in the Minnetonka refuge. So many birds were killed that the pheasant population has never been as abundant in that area since that time.

Pheasants are polygamous, and a rooster may have half a dozen or more hens in a harem. Males have leg spurs that can reach several inches long. They are furious fighters during mating season and have been accused of going into farm yards to challenge the local domestic rooster. The ring-neck's territorial call is a loud, harsh *kok-KACK*. Only the dominant rooster calls.

Hens make their nests of leaves and grasses on the ground and lay a clutch of 6 to 12 dark olive eggs. Incubation is about 24 days. The chicks will leave the nest in about an hour after hatching. In a week they can already take flight.

Winter blizzards can be devastating to pheasants. They tend to stand facing a storm, and the ice and snow that blows into their nostrils often suffocates them. They also find it difficult to burrow under the crusted snow for food or shelter. However, the destruction of critical habitat by humans is the main threat to pheasants, as it is to all wildlife.

In Japan, pheasants are said to feel tremors of imminent earthquakes and are considered an early warning system.

Winter

photo: John Toren

A Winter Walk

When the ominous purple snow clouds of November start rolling in, the landscape becomes a two-toned scene—drab brown and frigid white. But upon closer observation, we'll see that winter has a rich life of its own, and we can begin to enjoy elements of nature that have long been overshadowed by the riot of summer green.

For example, early winter is a great time to find and identify bird nests. The wispy thistledown nest of the American goldfinch can be found still securely bound in the crotch of a shrub or large bull thistle. The scraggly nest of the eastern kingbird hanging at the far end of a branch exposes the messy house builders that the kingbirds are. Who would guess that this gray and white tyrant would be the strongest defender of its home, chasing away hawks and crows many times its size? Flocks of American tree sparrows, identified by their rusty caps and the black stick-pin on their

Goldenrod galls

breasts, feed in the protection of the prairie grasses, and slate-colored juncos flit away flashing their gray and white tail feathers.

Chickadees and downy woodpeckers can be seen clinging to the stems of goldenrod, pecking at a swollen gall trying to get at the tiny morsel of white larva tucked away inside. The gall is a round swelling on the stem of the plant caused by the larva of a spotted-winged fly. The fly lays its eggs on the goldenrod stem in May and June. When the egg hatches, the larva burrows into the stem and the gall forms around it. It overwinters in the larval form inside the stem, hoping to escape the attention of chickadees and downy woodpeckers. If you see a hole in one of these galls, you can be sure a chickadee or woodpecker has had a tasty snack.

As late fall slides into early winter, the colors of the prairie grasses dominate the scene. The yellow-gold of the prairie cordgrass stands with the russet of big bluestem, its turkey-foot seedhead glistening in the weak sunlight of winter. Hairy awns of golden-brown Canada wild rye hang with their ripened seeds nodding as if in submission to the power of Father Frost.

The panicles of the switch grass look like little ancient umbrellas with only the spines remaining. Frosted clusters of little bluestem dot the late-fall prairie with their feathery bronze-orange spears.

Ragged stalks of curled dock and mullein stand like lonely sentries unwilling to abandon their posts. The faint smell of anise is still in the dried leaves of giant hyssop, and the scent of mint lingers in the leaves of wild bergamot. Monarch butterflies have migrated to their wintering grounds in Mexico, but their host plant, the common milkweed, remains behind, chewed and tattered with its nearly empty seed pods still releasing a few fluffy seeds as puffs of wind scatter them about.

Attached to the branch of a tree like a gray paper dirigible is the abandoned nest of the bald-faced hornet. If we look inside we can see six to eight quilted layers that maximize the insulating affect and provide insects and spiders a refuge from the bitter storms of winter.

Open scrapes on the ground and thrashed branches of a shrub or small tree are telltale signs of a white-tailed buck in rut. They let a passing doe know that a potential mate is in the neighborhood. In soft patches of snow, cottontail rabbit tracks can be seen followed by those of its pursuer, the red tailed fox. The footprints in the snow remind us of the timeless chase between predator and prey.

Short-tailed weasels, though not uncommon, are seldom seen since they hunt mainly at night. The black-tipped tail seems to be incongruous with the white body that helps camouflage the weasel in the snow. However, tests with fake weasels, with and without black-tipped tails, show that red-tailed hawk predators always captured all-white weasels and regularly missed those with black-tipped tails. The black tip on the tails apparently confuses the hawk, thus ensuring the weasel's survival.

Weasels are the ultimate mouse predators. Several years ago, while working in the sugarhouse on a frigid December day, I was visited by a short-tailed weasel. It was a dramatic sight! When it had finished inspecting the sugarhouse for mice, it stopped at my shoe and looked up at me as much as to say, "You're too big to eat!" and then scampered away.

If we take a winter walk, views and creatures like these are presented to us by Mother Nature free of charge. What a wonderful way to help warm the winter days and shorten the long winter nights!

photo: David Brislance

In winter the short-tailed weasel takes on a new coat and a new name: the ermine.

Cardinals

As the sun rises ever higher in the sky during the often frigid days of January and February, long stretches of glacial weather eventually give way to warming trends. And by the time the spring equinox arrives, signs of the coming season have long since begun to appear, and one of the first is the *what cheer, cheer, cheer* call of the northern cardinal. Unlike most bird species, both male and female cardinals sing. Males begin to sing in midwinter, and the females begin to answer as the mating season nears. Their persistent call can be heard from the first glints of light in the morning sky to the final gray shades of dusk.

photo: David Brislance

The expanding range of the cardinal in Minnesota is a recent development. Cardinals have been permanent residents in the bluff country south of Red Wing only since the 1940s. Backyard feeding stations are believed to have helped in their northward expansion. Nowadays cardinals can be seen overwintering as far north as Duluth.

The northern cardinal has two broods per year, with the female building the nest and incubating the eggs. The male takes part in feeding the young. Cardinals typically mate for life and maintain a territory ranging from 3 to 20 acres.

Cardinals molt in late summer, and the male's new feathers are tipped with gray. As winter progresses the tips wear off, producing a bright-red breeding color.

Cardinals are omnivorous, consuming a huge variety of plant and insect food, including caterpillars, grasshoppers, wild grapes, dogwood fruits, smartweeds, and sumac berries.

A frozen frog half-embedded in the mud.

Winter Survival Kit

Animals who must endure biting arctic winds and long, bone-chilling nights have developed a variety of means to adapt. For example, three frog species commonly found in Minnesota—the wood frog, the gray tree frog, and western chorus frog—crawl under leaf litter, rocks, or fallen logs to wait out the winter months. During that time 65 percent of a wood frog's body cavity and spaces between its cells may freeze, yet no ice crystals form within its cells, because these frogs can produce alcohol glycerol (anti-freeze), which protects the cell membranes, and glucose, which restricts ice formation outside the cells. Many insects also do this in anticipation of a freeze, but frogs wait for the ice to form on the skin before producing these chemicals, though they then become frost tolerant within a day.

When a frog begins to freeze, its body creates an alarm that shoots adrenaline into the bloodstream and activates the liver to produce a massive amount of glucose, essentially packing the frog's cells with anti-freeze. This massive dose of glucose would be enough to send us humans into a coma and death several times over, but in frogs, it acts as an agent to draw water from the cells, while also depressing the metabolic rate to conserve as much energy as possible. These frogs can survive with body temperatures as low as 19 degrees Fahrenheit for several weeks. When we hear the frogs singing in the spring, it's worthwhile reminding ourselves that they've been on a cryogenic nature trip that we can only imagine!

Before migratory patterns were understood, a common fable held that the swallows that skimmed over ponds every spring spent their winters hibernating in the mud under the ice. The idea of birds hibernating was thought to be ridiculous until a poorwill (a western cousin of our whippoorwill) was found in a comatose state in the Chuckwalla Mountains of California. Readings showed the bird's internal temperature to be the same as the air temperature, usually the case when an animal is dead. There was no detectable heartbeat or breathing, and its condition was identical to that of an animal in hibernation. When held in the hands, the bird would awaken and fly away. The Hopi Indians long before had understood the nature of this bird and called it "the sleeping one."

Black-capped chickadees bustle with feeding activity through the severest winter weather. While they do not hibernate, they shiver constantly at night and go into a torpor that lowers their body heat by 20 degrees Farenheit. Blood flow to their feet is constricted, giving the circulatory system enough time to reheat the blood before reentering the heart chamber. The feet stay warm, and the heart is not overstressed with frigid blood.

Chickadees have a denser layer of feathers than most other birds, and by puffing up and tucking their exposed eyes and beak into these feathers, they can keep heat loss to a minimum. Even with these adaptations it was found that the amount of their body fat dropped from 7 percent to 3 percent overnight. This might give us an idea of the large quantity of food they must consume each day just to survive the 15-hour winter nights. One way to help chickadees is to provide them with a roosting box where they can huddle together and be protected from the icy blasts of Old Man Winter.

photo: David Brislance

The poet may write that April is the "cruelest month," but for wildlife, January is the true test of survival.

photo: D'Arcy Norman

Winter Pond Walk

An early winter walk around a woodland pond is a relaxing way to engage our senses and restore our spirits. As I set out on such a walk, feathery snowflakes fill the air, collecting on the branches of red osier dogwood to add a Christmas-like theme to the trail. The fluffy flakes remind me of childhood days when I would stick out my tongue and scamper around trying to catch the frosty treats as they whirled by. A gust of wind rustles the dried leaves of a red oak tree that cling tenaciously to its branches. This cold burst of air reminds me that Boreas, the god of the north wind, is about to descend upon us, bringing another layer of winter cold.

A fallen basswood tree provides a comfortable resting spot as I sit, watch, and listen for activity nearby. A red squirrel suddenly scolds me from his perch on the branch of an oak tree. On the far end of the log I'm sitting on I see the chewed remains of acorn shells. This pile of leftovers is called "middens," and it's an indication that this feisty, incessant chatterer has made my resting spot his dining area. No wonder he's upset!

photo: David Brislance

A frisky red squirrel can chatter up a storm.

Along the hillside tom turkeys are noisily tossing leaf litter hither and thither as they scrounge for any morsels of food the squirrels and deer may have overlooked. The tufted beards dangling from their breasts look like little black dust brooms sweeping the ground. Their brilliant bronze and purple feathers are illuminated by the stream of sunbeams penetrating the leafless canopy.

A thin layer of ice has formed on the pond, and it now looks like a shimmering sheet of aluminum foil. Across the pond a flock of crows are raucously calling as they mob their mortal enemy, the great horned owl. I am amazed, once an individual crow sounds the alarm, how quickly its companions come to the rescue. Like firefighters racing to the scene of a blaze, they are true musketeers: "All for one and one for all."

Olive-green spears of horsetails along the frozen pond edges stand out among the muted browns and grays of December. Horsetails and their fern allies grew as giants 200 million years ago, and their fossilized remains created the coal deposits that now fuel our power plants. Horsetail is also called scouring rush because of the large amount of silica in its hollow stems, which made the plant useful to Native Americans and early settlers for cleaning dishes and utensils.

There are plenty of other sights and sounds to experience in the winter forest, along with the intervals of peace and quiet, but suddenly I've got something else on my mind: warming my hands on a cup of hot chocolate!

Watching Winter from a Window

The first snows of November can be heavy and slushy—perfect for children to romp in as they make their first Frosty the Snowman of the season. Unfortunately for those who prefer the warmer days of summer, there is no magic in Frosty's hat that will make the winter landscape green again.

Even on the coldest days, when Jack Frost has decorated the window panes with his arctic rime, I can still see much activity in my backyard. Gray squirrels are active even on the coldest days, fattening themselves from a feeder of whole corn and searching for any spillage under birdfeeders. Watching their persistent, acrobatic attempts to conquer the obstacles protecting the birdfeeders is like watching tightrope walkers at a circus. Any tree branch within six feet of a birdfeeder makes a perfect launching pad for these furry high-fliers.

Meanwhile, cottontails jump straight up in the air and chase each other through the new-fallen snow, delighted with the change of seasons. A doe and her two yearling fawns cautiously tip-toe through the yard, browsing on the shrubs we

conveniently provide for them. They have unknowingly become the bane of many gardeners, but their sleek brown coats and beautiful dark eyes brighten up the dreary, gray stratus cloud cover of early winter. When startled they prance away, waving bye-bye with their white tails.

A male cardinal perched on the bough of a snow-covered spruce tree appears on many a Christmas card. That splash of red satisfies a need for color. How much nicer it is to see one just outside the window! Meanwhile, hairy and downy woodpeckers hang upside down from suet feeders as they ravenously attack the calorie-laden squares of peanut cakes inside. The hairy is noticeably larger than the downy, and the two can also be differentiated by the size of their bills—the downy's is short and "cute," the hairy's is as long as his head.

Slate-colored juncos feed on the ground, scratching back and forth as they search for Niger thistle seeds in the melted snow under the feeders. (It's been estimated that there are more juncos than people in North America.) Blue jays alert one another when more corn is available and gobble up as many as seven kernels before flying off to cache their treasures in nearby hiding spots. Goldfinches and house finches are regulars at the feeding station but the unexpected arrival of pine siskins, redpolls, or a red-breasted nuthatch is a thrill that can warm the heart even on the most frigid winter days.

We need only to provide a feeding station and the nearby creatures will soon arrive, raising our spirits during the often gloomy days of winter.

photo: David Brislance

A red-breasted nuthatch comes in for a landing

photo: Carol Wahl

Coyotes are becoming a more common sight in suburban neighborhoods.

Coyotes

The mournful cry of the coyote reminds us of scenes from old Western movies, often with doom impending. When coyotes get together they often bark and yip together, creating a high-pitched, off-key chorus that fills the heart with mystery and wonder. They may be claiming territory, chasing prey, or merely socializing. Coyotes have more vocalizations than any other wild mammal in North America. The genus species name is *Canis latrans*, which means "barking dog." But a howling coyote sounds less like a dog than like a miserable banshee, and when a pack begins to howl together at sunset, hitting all the wrong high-pitched notes, it can be a hair-raising experience.

Coyotes are wary of humans and hunt mostly at night. But as coyotes expand into urban areas, they discover they're no longer being hunted and sometimes become less fearful of humans.

Coyotes are omnivores and eat roughly three pounds of food a day, including fruits and berries, with a special hankering for watermelons. They're considered a keystone species in the environment because they keep populations of mice, rabbits, raccoons, feral cats, and deer in check, thus reducing habitat degradation and increasing the likelihood that ground-nesting birds will breed successfully. Unlike packs of feral dogs who kill animals indiscriminately, coyotes kill only what they can eat.

The Blue Jay

Though there are other blue-colored jays in North America, the blue jay's iridescent blue feathers, spattered with black and white, and its ebony-feathered neck yoke make it easy to identify. In 1766, Jonathan Carver wrote, "The blue jay can scarcely be exceeded in beauty by the winged inhabitants of this or any other climate."

Blue jays are often seen in raucous flocks of six or more. The genus name translates as "blue chatterer." Their *toolool* courtship call is a welcome sign of early spring. They become very quiet and secretive when nesting, and they mate for life. Blue jays do rob other birds' nests, occasionally, but such predatory action is limited. One study found only 1 percent of a blue jay's diet consisted of bird material. Much of their diet consists of things we could all do without. They are major consumers of gypsy and tent moths, as well as hairy caterpillars, which few other birds consume.

Blue jays can hold up to six acorns in their throat pouch. They often bury the acrons, and sometimes neglect to retrieve them. It has been estimated that the great eastern oak forest has spread roughly 400 yards northward every year due to the blue jay's forgetfulness. Researchers even speculate that the size of an acorn is an evolutionary adaptation to the size of the blue jay's bill. Blue jays have, in turn, adapted to many habitats besides oak forests, and they can survive on a wide variety of food sources, including berries, food scraps, corn, weed seeds, and small invertebrates.

Though we often see blue jays in the winter, they do migrate. Northerly jays come south to replace the summer birds who abandon our backyards when the weather turns cold. Along the Great Lakes flocks of 50,000 have been seen migrating. They are expanding their range westward and have hybridized with their western cousin the Stellar's jay.

The Cherokees say that if you soak a blue jay's feather in water and gently brush it over a sleeping child's eyes, the child will be an early riser—a story many parents could appreciate! A bizarre bit of Southern folklore has it that the black ring around a blue jay's neck was created when a sparrow harnessed the jay to a plow.

The juniper (or red cedar) is one of the most widely distributed plants in the world.

Junipers

As a youngster growing up on a farm, one of the joys of my holiday season was going into the woods to find a Christmas tree. There was always an abundance of eastern red cedar (*Juniperus virginiana*) to choose from, and though the tree we selected may not have been perfectly formed, it brought the pungent smell of fresh-cut cedar into our living room. Once the tree was adorned with colorful lights, showy bulbs, and shimmering tinsel, we were very impressed and were sure Santa would be, too.

Eastern red cedar, also known as juniper, was once the holiday tree of choice, but it has long since been replaced by several species of fir and pine. In the wild, junipers can become invasive, especially on dry, limestone bluffs where they crowd out native wildflowers. They do, however, provide food for at least 90 species of birds as well as nesting sites and protection from the elements.

Cedar waxwings were so named because of their affinity for the blue berries of the juniper. The fruit that passes through their digestive system—it takes 12 minutes—is three times more likely to germinate than berries that simply fall to the ground. Cedar waxwings and other fruit-eating birds like robins and bluebirds often sit on fences and power lines, which explains why we so often see junipers growing beneath these structures.

The tree is rot resistant and has been widely used for fence posts, chests, and closet linings—moths don't like the scent of the wood. And juniper berries are still used to flavor gin.

The Cherokee and other tribes consider the cedar to be the "tree of life" containing the spirits of their ancestors. It was burned during ritual purification ceremonies. In Europe, the *white* cedar was given the name arbor vitae (tree of life), though red cedars (junipers) were planted in front of houses to keep witches away. It was believed a witch had to count every needle on the tree before she could enter the house—an impossible task. Our tree was used to please jolly old St. Nicholas so he would enter our house with his bag of toys and goodies—no needle-counting required!

I was given a cedar pencil box for my first communion and still open it on occasion just to enjoy the residual scent of cedar. It is in my nature memory book.

Juncos

photo: David Brislance

In referring to the dark-eyed juncos, Henry David Thoreau said, "They come over the ground like snowflakes before the northeast wind"—a poetic way of describing the southward migration of juncos, which comes as a prelude to winter. These members of the sparrow family arrive in Minnesota in early October and remain with us throughout the bitter, sun-starved days. The species name for the junco is *hymalis* which in Latin means "of the winter." Juncos have white breast feathers and slate-gray backs, resembling winter itself, with its leaden skies above and snow below. Often unnoticed as they scour the underbrush, juncos become easy to see and identify as they burst into flight, flashing tails edged in white. During the mating season males show off their white finery in hopes of impressing nearby females.

In a movement called differential migration, adult females migrate farther south than young females; young males remain farthest north, with adult males in the middle between these other groups. Junco flocks of 15 to 25 are common around backyard feeders. Their distinctive habit of double scratching the ground as they search for seeds is fun to watch.

In the mild days of April, when mud puddles form and maple sap drips, juncos gather in flocks of a hundred or more preparing for the return flight to their northern breeding grounds. Their dainty, tinkling trill is a musical finale as they leave us until the nipping air descends again.

The Gray Squirrel

Gray squirrels are often formidable adversaries, pesky and innovative, at our backyard birdfeeders. Their feisty never-say-die antics can provide us with hours of entertainment, but they can also raise our blood pressure. These relatives of Rocket J. Squirrel do not consider a distance of 10 feet or more from a nearby limb to the feeder too daunting a leap. And though attempts to outwit a pole collar or a counterweight often end in failure, gray squirrels seldom give in easily. If at first you don't succeed, try and try again.

The squirrels in my yard help themselves to the whole corn I put out for the blue jays and overwintering mallards, and they've developed fat little bellies as a result. Come spring, a nice crop of corn emerges in the lawn due to the industrious work of these furry friends, who are better at burying seeds than finding them again later.

The average gray squirrel weighs a little over a pound (mine are the exception) with an eight-inch tail. Squirrels shed their hair twice a year except for their tail, which loses hair only once a year. The genus name *Sciurus* comes from the Greek meaning, "animal that sits in the shadow of its tail."

A gray squirrel builds its nest, which is called a "drey," in the fork of a tree out of twigs and dried leaves. These dreys are easy to spot in the wintertime, when many trees are otherwise bare. Females will move their young often from nest to nest to avoid predators and to get away from flea infestations, which pose a serious health

A few of the albino squirrels of Olney, Illinois

problem for squirrels. All the same, about three-quarters of newborn squirrels die the first year due to heavy predation from snakes, hawks, owls, foxes, bobcats, weasels, raccoons, and even cats and dogs.

A gray squirrel will cache several thousand nuts and acorns in a season. Hickory nuts, butternuts, and walnuts must be buried in order for them to stay moist; squirrels have a keen sense of smell, yet they never relocate all of their stashed nuts, and the ones they overlook play a key role in the survival of these species. For example, about 95 percent of all hickory trees come from nuts buried by squirrels.

Black squirrels are a color morph of the gray variety. Black squirrels often become the dominant color in a breeding population. In my home town of Wabasha, black squirrels are more numerous than grays. I remember seeing my first black squirrel in the woods in the late 1970s and thinking it was a new species. In Olney, Illinois, albino squirrels are protected and have the right of way on streets and sidewalks, though dogs and cats are not allowed to run free. Anyone running over a white squirrel can be fined 750 dollars.

Cinderella actually wore squirrel slippers but the French word got mistranslated as glass slippers. Squirrel slippers sound more comfortable.

In some Native American tribal myths, squirrels are considered gossip mongers and troublemakers. However, other tribes consider them to be the guardians of the forests and admire their food-gathering skills and courage. These contradictory images of the squirrel are shared by many suburban homeowners even today.

photo: David Brislance

A downy woodpecker alights on a branch.

Woodpeckers

Woodpeckers are the percussionists of the forest They can execute the familiar rat-a-tat-tat at a speed of 15 miles per hour, and keep it up at a pace averaging 12,000 times a day.

How can a woodpecker's head withstand all the hammering? Their unusual adaptations are numerous. Woodpeckers have a spongy bone as part of their skull that acts as a shock absorber to the brain, and their neck vertebrae are more flexible than those of other birds, relieving pressure during drumming. When a woodpecker pecks at an object it doesn't twist its neck, which further reduces the possibility of injury. Woodpeckers also take a break now and then as they drum, allowing the brain to cool. They have nictitating eyelids, which act as a safety belt to keeps their eyes from popping out of their heads while they're hammering, and also protects the eyes from flying pieces of woodchips and sawdust.

Woodpeckers do more to preserve the health of trees than any other species. They remove injurious insects and larvae from the bark and usually peck into trees that are already diseased. The yellow-bellied sapsucker, a member of the woodpecker family, can damage healthy trees but also provides an early spring food source for butterflies and other birds. The ruby-throated hummingbird relies on the sapsucker for food in early springtime when few flowers are blooming. Ruby-throated hummingbirds are thought to extend their northern range only as far as the yellow-bellied sapsucker.

Woodpeckers were once called lightning birds because it was believed their red feathers protected people from lightning strikes. California tribes used red feathers for currency, and the Omaha tribe traditionally believed the woodpeckers were a protector of the young because they kept their own offspring in safe places. The red feather pattern on the heads of male downy woodpeckers is unique to each individual, like the fingerprints of humans.

When woodpeckers come knocking on your door, don't forget all of the good things they do.

The handsome red-bellied woodpecker (right) is probably as poorly named as any bird. Everyone thrills to the sudden appearance of the pileated woodpecker (above).

photo: David Brislance

The Horned Lark

In February and early March, when winter seems to be loosening its grip on sunny days, it's not uncommon to see a flock of light-colored, upright birds scatter from a country road as a vehicle approaches. It's likely that they're horned larks. If you happen to get a closer look, you'll see the bird's black sideburns, yellow throat, and maybe even the two feathered tufts on its crown from which it derives its name.

The horned lark is the first migrant to return each spring, and once it arrives, it refuses to be driven back by frigid temperatures or wintery blasts. Horned larks will have two or three broods a year, though they sometimes lose the first one due to brutal March weather. At that time of year females have been seen removing frozen eggs and laying new ones. Larks build their nests by scratching a cavity into the ground and filling it with grasses and pieces of corn stalk. The female will place a pebble, corncob, or other material on the "doorstep" of the nest to hide the pile of dirt she created while excavating it. Males will fly 500 to 800 feet in the air when mating and sing while they descend.

The family name of the lark is Alaudidae, which derives from the Latin word for lark, *alauda.* The Roman naturalist Pliny suspected the word had a Celtic origin. Percy Bysshe Shelley, in his poem "To a Skylark," penned "Hail to thee, blithe spirit"— a fitting tribute to our first spring migrant.

Holiday Gifts

The natural world provides us with many gifts during the holiday season. A cardinal on a snow-covered bough with his glowing red feathers set against the sparkling snowflakes is nature's Christmas card—sent with no postage necessary.

Copper-colored oak leaves hanging stubbornly to a branch in the breeze are temporarily silenced by a visit from the frost fairy during the cool night hours. The amber rays of the early morning sun illuminate the hoary leaves, keenly displaying the handiwork of this forest pixie.

This is the time of the Rutting Moon. White-tailed bucks are dressed in their auburn winter coats and adorned with splendid pearl-colored antlers. They can bound effortlessly through the deep snow, almost taking flight like Santa's reindeer. Turkeys softly cluck as they pluck at acorns and other edibles on the frozen ground. Their rummaging results in indiscriminate piles of leaf litter that resemble the mess of a children's Christmas party after all the presents have been opened.

A cottontail darts from a brush pile along a well-worn path, suggesting that this is its home-sweet-home. Extensive chew marks on the bent branches of a nearby red osier dogwood identify its favorite place to dine.

A barred owl sits stoically in the hollow of a sugar maple tree with its eyes closed, seemingly taking a nap as it digests some unfortunate critter that it chased down in the forest the previous night. However, the bird is wary of any human intrusion, and when approached it will fly off without a sound deeper into the forest, using the same stealth that makes it such a good nocturnal hunter.

These gifts are just a few of the many that nature makes available to us throughout the winter months. All we need to do is remove ourselves for an hour from the bustle of human activity, step outside, and take a look.

Snow Birds

By the time winter snows hit the region, many of our bird species have long since departed for warmer southern climates, but a few remain, and the cold winter weather also encourages more northerly species to push south into our area. Cedar waxwings can often be seen feeding on juniper berries, and flocks of robins overwinter where there's food and open water. Crabapples are a favorite food source, and as long as the food supply lasts, some robins will stick around.

The juncos usually arrive in early October, reminding us that Old Man Winter is approaching. With their slate-colored backs and white undersides, juncos blend nicely into the gray skies and late autumn landscape. In early April, their tinkling song resonates through the shrubs as they bid us farewell.

Another bird species that enters our area for the winter is the American tree sparrow. It can be identified by its rusty rufous crown and the black spot on its chest. Tree sparrows are often found in flocks in tall grassy areas. Their soft jingling call and sharp *tsiiw* chip can be heard as they furtively flit among the vegetation feeding on the seeds of ragweed, crabgrass, lambs quarters, and pigweed.

Four more exotic species that come south to our region in winter are the snow bunting, evening grosbeak, red crossbill, and common redpoll. Of these the snow bunting is undoubtedly the most common. I sometimes spot a few individuals in early November in the open fields at the U of M Horticultural Research Center. As winter advances, flocks of several hundred sometimes appear in southern Minnesota. Their winter plumage is white and rusty brown, but when flocks are in flight they look like a whirling snow flurry—thus their other common name, "snowflake."

Snow buntings nest on the ground in subarctic regions, roosting beneath a weed, bush, tuft of grass, or a burrow of soft snow to protect themselves from the bitter winter nights. The winter diet of these "flying snowballs" consists of the same weed seeds that the tree sparrow enjoys.

Evening grosbeaks are the most tame of these plumed visitors, often allowing humans to get within a few inches before flitting away to a nearby branch. The male, with its dull yellow plumage and black-and-white wings, looks like a plump, overgrown goldfinch. The female is silvery. Its large beak is the bird's most distinguishing feature—hence the name. The Ojibwe call it *paushkundamo*, the breaker of soft fleshy substances.

The common redpoll is a little, streaked, gray-brown finch, easily distinguished from sparrows by the big red spot on its forehead and black chin. The breast of the male is tinted a pale pink.

The red crossbill is about the size of a house sparrow. The male is dull red, and the female is olive-gray with a yellowish rump and underparts. The distinguishing feature of the crossbill is its crossed mandibles, which are effective in clipping off the scales and extricating the seeds from the cones of spruce and other evergreens. Crossbills are adept at climbing acrobatically through trees like little parrots.

One of the most important adaptations enabling redpolls and crossbills to withstand the great energy demands imposed by our winters is a bi-lobed pocket midway down the neck. Called an esophageal diverticulum, this pocket serves as a storage place for seeds toward nightfall and during severe weather. Redpolls and crossbills seek protection in dense evergreen stands, fluffing up their feathers to retain as much body heat as possible. Studies suggest that redpolls can survive extreme temperatures better than any other songbird.

photos: David Brislance

(left) A red crossbill; (right) a common redpoll

Acknowledgments

I would like to thank my family and friends who have shared their experiences and appreciation of the natural world with me over the years. Thank you to the staff at the University of Minnesota Landscape Arboretum for sharing their knowledge and friendship, especially Julia Bohnen and Rich Gjertsen. Thanks also to Peter Olin, director emeritus of the Arboretum, and present Director Peter Moe, for their support and understanding.

A special thank you to Julie Brophy and Karen Sowizral, who have volunteered their time and energy to help make my walks and talks fun and enjoyable, and also to Norton Stillman, the publisher of Nodin Press, who encouraged me to write this book.

While I have written the material, the credit for putting it into book form goes to the editor, John Toren. John's knowledge and expertise have made this work possible and have given me the understanding that a great editor is the key to completing a book. How fortunate I have been to have John's help throughout this process.

I would also like to thank Jim Gilbert, whose love of nature has been an inspiration to me and thousands of other people.

Finally and most importantly, I would like to thank my wife, Susan Mauren, who has put up with me for 40 years and whose critique and help with this book were absolutely essential.

About the Author: Matt Schuth grew up on a farm near Wabasha, Minnesota, on the bluffs overlooking the Mississippi. He has worked in various capacities at the University of Minnesota Landscape Arboretum since 1982, managing maple syrup production, working on wetland projects with graduate students, helping to prepare and maintain research plots, setting up bluebird trails, wood duck houses, and osprey nests, and keeping track of bird, mammal, and butterfly sightings at the Arboretum. He continues to write about the natural world for the *Arboretum Magazine* and to conduct nature walks and talks throughout the year.